W9-CLZ-496

AFFORDABLE HOUSING AND URBAN REDEVELOPMENT IN THE UNITED STATES

URBAN AFFAIRS ANNUAL REVIEW

A semiannual series of reference volumes discussing programs, policies, and current developments in all areas of concern to urban specialists.

SERIES EDITORS

The **Urban Affairs Annual Review** presents original theoretical, normative, and empirical work on urban issues and problems in volumes published on an annual or bi-annual basis. The objective is to encourage critical thinking and effective practice by bringing together interdisciplinary perspectives on a common urban theme. Research that links theoretical, empirical, and policy approaches and that draws on comparative analyses offers the most promise for bridging disciplinary boundaries and contributing to these broad objectives. With the help of an international advisory board, the editors will invite and review proposals for **Urban Affairs Annual Review** volumes that incorporate these objectives. The aim is to ensure that the **Urban Affairs Annual Review** remains in the forefront of urban research and practice by providing thoughtful, timely analyses of cross-cutting issues for an audience of scholars, students, and practitioners working on urban concerns throughout the world.

INTERNATIONAL EDITORIAL ADVISORY BOARD

RECENT VOLUMES

23 **CITIES IN THE 21st CENTURY**
Edited by Gary Gappert and Richard V. Knight

24 **THE GREAT HOUSING EXPERIMENT**
Edited by Joseph Friedman and Daniel H. Weinberg

25 **CITIES AND SICKNESS: Health Care in Urban America**
Edited by Ann Lennarson Greer and Scott Greer

26 **CITIES IN TRANSFORMATION: Class, Capital, and the State**
Edited by Michael Peter Smith

27 **URBAN ECONOMIC DEVELOPMENT**
Edited by Richard D. Bingham and John P. Blair

28 **HIGH TECHNOLOGY, SPACE, AND SOCIETY**
Edited by Manuel Castells

29 **URBAN ETHNICITY IN THE UNITED STATES: New Immigrants and Old Minorities**
Edited by Lionel Maldonado and Joan Moore

30 **CITIES IN STRESS: A New Look at the Urban Crisis**
Edited by M. Gottdiener

31 **THE FUTURE OF WINTER CITIES**
Edited by Gary Gappert

32 **DIVIDED NEIGHBORHOODS: Changing Patterns of Racial Segregation**
Edited by Gary A. Tobin

33 **PUBLIC INFRASTRUCTURE PLANNING AND MANAGEMENT**
Edited by Jay M. Stein

34 **ECONOMIC RESTRUCTURING AND POLITICAL RESPONSE**
Edited by Robert A. Beauregard

35 **CITIES IN A GLOBAL SOCIETY**
Edited by Richard V. Knight and Gary Gappert

36 **GOVERNMENT AND HOUSING: Developments in Seven Countries**
Edited by Willem van Vliet–– and Jan van Weesep

37 **LEADERSHIP AND URBAN REGENERATION: Cities in North America and Europe**
Edited by Dennis Judd and Michael Parkinson

38 **BIG CITY POLITICS IN TRANSITION**
Edited by H. V. Savitch and John Clayton Thomas

39 **URBAN LIFE IN TRANSITION**
Edited by M. Gottdiener and Chris G. Pickvance

40 **THE PENTAGON AND THE CITIES**
Edited by Andrew Kirby

41 **MOBILIZING THE COMMUNITY: Local Politics in the Era of the Global City**
Edited by Robert Fisher and Joseph Kling

42 **GENDER IN URBAN RESEARCH**
Edited by Judith A. Garber and Robyne S. Turner

43 **BUILDING THE PUBLIC CITY: The Politics, Governance, and Finance of Public Infrastructure**
Edited by David C. Perry

44 **NORTH AMERICAN CITIES AND THE GLOBAL ECONOMY: Challenges and Opportunities**
Edited by Peter Karl Kresl and Gary Gappert

45 **REGIONAL POLITICS: America in a Post-City Age**
Edited by H. V. Savitch and Ronald K. Vogel

46 **AFFORDABLE HOUSING AND URBAN REDEVELOPMENT IN THE UNITED STATES**
Edited by Willem van Vliet--

AFFORDABLE HOUSING AND URBAN REDEVELOPMENT IN THE UNITED STATES

◆

EDITED BY
WILLEM VAN VLIET--

URBAN
AFFAIRS
ANNUAL
REVIEWS
46

SAGE Publications
International Educational and Professional Publisher
Thousand Oaks London New Delhi

For information address:

SAGE Publications, Inc.
2455 Teller Road
Thousand Oaks, California 91320
E-mail: order@sagepub.com

SAGE Publications Ltd.
6 Bonhill Street
London EC2A 4PU
United Kingdom

SAGE Publications India Pvt. Ltd.
M-32 Market
Greater Kailash I
New Delhi 110 048 India

Printed in the United States of America

Library of Congress Cataloging-in-Publication Data

ISSN: 0083-4688
ISBN 0-8039-7050-1 (Hardcover)
ISBN 0-8039-7051-X (Paperback)

97 98 99 00 01 02 03 10 9 8 7 6 5 4 3 2 1

Production Editor: Diana E. Axelsen
Typesetter/Designer: Marion S. Warren

Contents

Foreword

AVIS C. VIDAL

A generation ago, the nation's housing problems were broadly understood in terms of inadequate supply. Too many housing units were substandard, and too many households lived in housing that was deemed overcrowded. Faced with these problems, the desired public sector response was clear: Stimulate production. A variety of direct production subsidy programs administered by the U.S. Department of Housing and Urban Development (HUD) and an aggressively supportive tax code made the construction of assisted housing a relatively straightforward and lucrative activity, and private builders responded accordingly.

Much has changed. Today's housing problem is framed principally in demand terms: housing affordability. The federal policy shift toward demand-side subsidies—the demise of the Section 8 construction programs and increased reliance on housing vouchers—reflects that altered understanding of the problem (as well as a fundamental shift in the political climate).

Vouchers serve some households very well, specifically, those who are lucky enough to receive vouchers and whose demographic characteristics lead them to be well served by the private market. That does not include large families, particularly those headed by a single parent, and households of color, who continue to be subject to considerable housing market discrimination. And it does not include the substantial numbers of households who remain unassisted and for whom the supply of low-cost housing has dwindled steadily along with the stimuli to its production and operation.

The concentration of these households in weakening central-city and inner-suburban neighborhoods fosters problems shared by the entire community. These are problems whose amelioration goes beyond the realm of housing understood in simple bricks-and-mortar terms. They pose more demanding challenges than the problems of the past and require more complex and nuanced policy responses.

The essays in this volume speak to those policy challenges. They strike a welcome note of optimism at a time when many housing analysts and urban policy advocates are reaching for crying towels. And well they might. Many factors bode ill for poor urban communities as the decade draws to a close. The macro forces that have shaped the character of those communities—the out-migration of middle-class families, high levels of racial segregation, and the decline in economic opportunities, especially for low-skilled workers—remain powerful, and countervailing trends are in short supply. The political standing of cities and of the poor in Congress and in state legislatures is weak and seems likely to remain so for the foreseeable future. The national mood is ungenerous.

Current trends in public policy generated by this environment signal intensified challenges for those committed to strengthening urban communities and to assuring decent living environments for residents of limited means. Four are of particular relevance to the themes struck by these authors.

First, the nation's political shift to the right has gone hand in hand with rising public frustration with government and skepticism about the public sector's ability to make a positive difference in people's lives. The strategies discussed here argue otherwise. They tell the stories of effective action, demonstrating that collectively we have the capacity to devise workable programs to address the housing needs of low-income people and the problems of distressed neighborhoods. More fundamentally, they argue that we have the capacity to learn from and correct our mistakes, and hence that we have the capacity to respond constructively when old policies and practices collide with new events and circumstances.

Second, the public resources available to address the problems of distressed neighborhoods in central cities and older inner suburbs will be in short supply. Many cities face the prospect of severe resource constraints. Continued declines in federal support for housing and community development are likely, and their effect will be accentuated if accompanied by the also likely reduction of direct income and in-kind support for poor households. Cities in states that are themselves facing revenue shortfalls will be even harder hit.

This prospect has dual import for readers of this volume. On the one hand, it gives cause for concern because so many of the authors emphasize the critical role that federal dollars have played in crafting many of the successes described. On the other hand, the lessons the essays teach about how to spend housing and community development dollars wisely—and about how not to spend them—are an important resource in an environment where funds are tight.

Third, more decisions about how to spend declining federal housing and community development dollars will be made by states and localities. HUD's decision to require communities to prepare Comprehensive Plans showing how they will spend HUD funds from a variety of programs (rather than the simpler Comprehensive Housing Affordability Strategy, or CHAS) is clearly a harbinger of more to come—regardless of how the agency is ultimately "reinvented." Devolution will expand opportunities for the kinds of local innovations documented in the following pages, but will strain the capabilities of many local agencies that have yet to acquire the skills needed to negotiate the programmatic complexity of housing and community development in today's climate. Despite long-standing complaints about "Washington red tape," many mayors have quickly come to the realization that their agencies lack the capacity to deal well with the new responsibilities they seem destined to inherit. The mistakes and successes of jurisdictions with a significant track record, documented by these authors, provide valuable guidance for the growing ranks of public servants who will be newly challenged by the further devolution of responsibility from Washington.

Finally, a confluence of forces seems likely to expand and intensify the importance of collaborative program delivery. The remaining subsidized housing production programs, principally the low-income housing tax credit, make production astonishingly complex and yield housing that is far more vulnerable to market forces than the assisted housing of the past. The shallow nature of the subsidy and the perceived level of risk necessitate the participation of multiple sources of financing, including subsidies other than the credit itself; the result is so-called patchwork financing created by a sizable number of participants. Declining subsidy dollars will make these deals more challenging to assemble and will create pressure to further increase the number of "partners." The general unwillingness of private developers to produce such housing in central-city and inner suburban markets will continue to bolster the rise of the nonprofit housing sector. The increasing awareness that in such markets successful delivery of housing services requires a broad strategy that supports families and communities will have the same effect. The result: A growing number of localities in which new and stronger links among the public, private, and nonprofit sectors will be needed.

The essays assembled here explore the interaction of these factors in rich detail. They illustrate the great variety of forms that collaborations can take and the range of roles that nonprofits and the public sector can assume. In so doing, they help chart the course for those who understand that a critical task in the years ahead is the reconstruction of working relationships that integrate poor and at-risk neighborhoods and their residents into the broader community and economy. The opportunities for rebuilding this fragile "social capital," found throughout the volume, are possibly its most fundamental contribution.

Preface

In 1949, the U.S. Congress adopted legislation that called for "a decent home and suitable living environment for every American family." Nearly 50 years later, this goal remains an elusive dream for many. For numerous households still today, the reality is a daily struggle to overcome problems of housing and deficient neighborhood support structures. Although there is no agreement on precisely how many homeless people there are, there is agreement that their number is large and increasing. Many more households are experiencing housing affordability problems. This is evidenced by, for example, the continuing decline in home ownership rates among particularly the younger age groups and the excessive proportion of income paid for housing by many low-income renters. These and other housing problems do not occur in isolation. They are compounded by their frequent occurrence in racially segregated areas with inadequate municipal services; a defective physical infrastructure; an insufficient tax base; a scarcity of community institutions; high rates of poverty, unemployment, and crime; lack of medical care; and unsatisfactory access to quality education.

In the recent period, the federal government has shifted a significant part of the responsibility it previously assumed for addressing these problems to the states and local communities as well as to the private and nonprofit sectors. These trends of decentralization and privatization continue into the present. In response, cities across the country have developed different approaches to the new and complex challenges facing them. This book is an attempt to learn from their experiences. Its intent is to derive lessons about what works and what does not. It identifies ingredients of success as a basis for formulating guidelines that may inform current and prospective policies. In this connection, it examines questions regarding the transferability of what produces success in any particular place.

Just as there is diversity of viewpoints among those making practical decisions about how to deal with problems of housing and urban development, so also is there a range of perspectives among those who study and write about these problems. This book reflects some of that variation. For example, the case studies report on and evaluate various forms of housing management, including privatized public housing, tenant cooperatives, and resident control in mutual housing associations. Similarly, although most chapters provide dispassionate analysis, a few take a more active stance of advocacy on behalf of a particular approach.

Notwithstanding this diversity, several common themes present themselves, which consistently recur as elements contributing to successful outcomes. For example, effective approaches are comprehensive rather than sectoral, target a whole community rather than a certain population category, build strong coalitions with multiple partners, rely on the professional expertise of well-qualified staff, and involve participatory processes with meaningful resident involvement. The Foreword, the introductions to Parts I, II, and III, and the concluding chapter elaborate on these and other themes. They sketch out the background for the local situations and place them in their national context.

Finally, it is important to recognize that this book is primarily oriented to older and often declining inner-city neighborhoods. It is here where the nation's poor are increasingly concentrated. This focus is not to deny or belittle the existence of housing and development problems in other locales. It is merely an acknowledgment that the concerns and ameliorative strategies are different and that the preservation of available housing and the rehabilitation of older neighborhoods tend to take precedence over new construction and development.

Acknowledgments

The chapters that follow were invited and purposely prepared for this book. The authors have been accommodating in following guidelines that asked them not only to address certain substantive issues but also to examine specifically what one might learn from their analyses. They were also responsive to the comments that reviewers made on early drafts. This final version owes considerably to the discerning observations, thoughtful suggestions, and otherwise helpful contributions of Chuck Ackerman, Thomas Bledsoe, Rochelle Brown, Janice Cogger, Charles E. Connerly, Sara Eilers, Daniel Grulich, Lenneal Henderson, Ella Johnston, Langley Keyes, Edward Landon, Mark McDermott, Bart McDonough, Douglas S. Massey, Kate Monter, Mary Nenno, William Peterman, Robert Pyne, Donald Rapp, Karen Safer, Sola Seriki, Mara Sidney, Meg Slifcak, Dennis Taylor, Greg Van Ryzin, Chris Warren, Eleanor White, Bill Whitney, Bob Wolf, and Jordan Yin. Mara Sidney prepared the index with consummate skill, while at Sage, Diana Axelsen and Catherine Rossbach shepherded the manuscript through various stages with vigilance, tolerance, and humor. Finally, Susan Clarke and Dennis Judd, series editors of the **Urban Affairs Annual Reviews,** provided constructive criticism and collegial support without which this project would have hung a U-ie.

1 Changes and Challenges in Affordable Housing and Urban Development

MARY K. NENNO

Abstract

The way in which affordable housing and urban development activity is carried out in the United States is changing. Part of this change involves the constantly shifting pattern of assistance provided by the federal government in terms of funding levels, the wide array of programs, assistance methods, and delivery mechanisms. It also concerns the uncertain mission of the U.S. Department of Housing and Urban Development. In counterpoint to the federal government's role, there is the growth over the last decade of a new capacity at state and local levels to finance and carry out such activity with a range of new partners, often without federal assistance or requirements. A final factor influencing the direction of affordable housing and urban development is the changing pattern and structure of urban life itself, in particular, the economic structure, including the joining of interests between central cities and their metropolitan areas and the increasing recognition of the urban region as the focus for action in the future. All these changes are in flux with no settled resolution. But it is clear that new relationships and more comprehensive strategies will be required to link the resources of all levels of government with the private sector and with the growing capacity of neighborhood citizen organizations, if the changing needs of urban America are to be addressed.

■

Meeting the changes and challenges in affordable housing and urban development requires attention to the following major areas: the structure of programs and assistance of the federal government; the emerging capacity at state and local levels; an understanding of the ways in which urban areas are changing; and more effective strategies to link and target resources toward specific goals. This book identifies and describes initiatives to restore blighted urban areas and severely distressed public housing developments. It offers a new understanding of the forces that are required to achieve success in revitalization and provides lessons for future efforts.

The Unsettled Pattern of Federal Assistance

Affordable housing and urban development assistance by the federal government is currently diffused and unfocused. The lack of a clear mission by the U.S. Department of Housing and Urban Development (HUD) and short-term, incremental actions by the Congress in authorizing and funding these activities have resulted in an array of HUD-administered programs. The 1994 analysis of HUD organization and management by the National Academy of Public Administration identified some 200 programs accumulated over time; the number of HUD programs grew from 54 in 1980, although the 1980s was also a period of declining budgets and staff. In the words of this report, "This program overload saps HUD resources, fragments the Department's work force, creates impossible expectations, and confuses communities" (National Academy of Public Administration, 1994). Despite this uncoordinated growth, two new programs were proposed by the Administration and approved by the Congress in 1993: Empowerment Zones and Enterprise Communities. In its fiscal year (FY) 1995 budget proposal, HUD recommended 15 new programs as noted by the Subcommittee on Housing and Community Development of the House of Representatives (Roukema, 1994). The Congress did not pass a housing authorization bill in 1994, thus most of these new proposals were not enacted. In any event, the Congressional elections in November created an entirely new political climate for housing and urban development programs. In mid-November, the Department accelerated its timetable to "re-invent" HUD by proposing the consolidation of all HUD programs into three comprehensive block grants: (a) an Affordable Housing Fund built on the HOME program to cover new housing construction/rehabilitation (b) The Community Opportunity Program that consolidated the Community Development Block Grant Program (CDBG) and other economic development initiatives; and (c) Housing Certificates and Vouchers for Families and Individuals that would eliminate the project-oriented public housing and Section 202 elderly and handicapped prodid this happen?

grams. These proposals were incorporated into the Department's FY 1996 budget proposal presented to the Congress in February, 1995. They require authorization by the newly elected, Republican-controlled Congress. Such authorization is uncertain. In February, 1995, the House of Representatives proposed rescinding $7.2 billion in HUD's approved appropriations for FY 1995. In 1996, the debate over the structure of HUD programs continued; there is certain to be a period of uncertainty and confusion covering both funding and future direction.

One question raised periodically is the appropriate role of government in the provision of affordable housing; a few have even questioned if there *is* a public role. This is so despite the long tradition, dating back to the 1930s of the public role, and despite experience over the past decade documenting that both the private and nonprofit housing sponsors depend on government support, in particular, the use of Community Development Block Grants (CDBG), if they are to achieve housing affordable to lower-income families. In terms of the direct role of government in constructing and management of lower-income housing, there is not only tradition but a convincing rationale that a local public entity is important in filling in gaps in coverage left by private and nonprofit housing sponsors and maintaining a stable entity over time that can respond to urgent lower-income housing needs. As indicated below in the section on "Emerging State and Local Capacity," traditional functions of general purpose local government (including such functions as building codes, zoning and land use regulations, minimum housing standards codes, and real estate taxation) have important impacts on housing development. In addition, as also documented in this section, the new government initiatives taken in the decade of the 1980s (including public-private partnerships, low-income housing trust funds with dedicated public revenues, issuance of tax-exempt bonds for housing development, impact fees on commercial development for housing purposes, inclusionary zoning, density bonuses, housing rehabilitation loans and grants, and tax increment financing) all provide important resources for a variety of housing entities to carry out lower-income housing initiatives. There is a basic public role in achieving affordable housing.

pub. role in housing

The Shifts in Housing Assistance Programs

The constant shift in HUD's housing assistance programs is documented in the fact that of the currently existing programs only public housing (59 years), Section 202 elderly and handicapped housing (37 years), and Section 8 Certificates/Vouchers (22 years) have long-term records. Excluding the Section 235 lower-income homeownership program (terminated in 1989), the remaining six programs have an average life span of less than seven years (see Table 1.1). In addition, the annual level of HUD-assisted housing units put in place has steep peaks and valleys with new construction-substantial rehabilitation starts ranging from a low of 4,916 housing units in 1956 to a high of 309,982 housing units in 1970. The annual level of newly authorized incremental housing units has been below 100,000 units since 1988 (U.S. Department of Housing and Urban Development Budget Summaries, FY 1988 to FY 1995). The appropriation for newly authorized housing units for FY 1995 is $11.1 billion up from levels as low as $7 billion in the late 1980s, but well below $30 billion in 1980; a substantial amount of this is proposed to be rescinded, as noted above. In addition, there has been a dramatic shift since the late 1970s away from new construction of low-income housing to the leasing of existing housing under the Section 8 Certificate/Voucher program, reaching a high of 87% in 1978. In HUD's FY 1995 budget, the new construction component in low-income housing was restored to a level of 60%, largely reflecting the implementation of the HOME assistance program adopted in 1990; whether this level will be sustained is not clear, because certificates/vouchers are less costly in the short run than newly constructed units. The recent heavy emphasis on the leasing of private housing has raised a policy issue concerning whether subsidizing "income support" to help low-income families afford available housing is an appropriate role for the Department of HUD, rather than its traditional role of improving the supply and condition of housing. Income support is usually viewed as a priority function of the Department of Health and Human Services or more recently, under Low-Income Working Family Tax Credits.

■ The Changing Shape of Urban Development Assistance

While the HUD functions of urban development and redevelopment have shown less volatility than those of affordable housing, they also have been subject to change. The Urban Renewal program was replaced by the Community Development Block Grant (CDBG) program in 1974 and has moved away from long-term redevelopment activity to shorter-term

TABLE 1.1 The Cycle of Federally Assisted Housing Programs

Program	*Year Enacted*	*Terminated*	*Years in Operation*
Public housing	1937	—	59
Section 202 elderly or handicapped	1959	—	37
Section 221-d-3 rental	1961	1968	7
Rent supplements	1965	1974	9
Section 235 homeownership	1968	1989	21
Section 236 rental housing	1968	1974	6
Section 8 new construction	1974	1983	9
Section 8 existing housing	1974	—	22
Housing action grants (HODAG)	1983	1989	6
HOME program (block grants)	1990	—	6

SOURCE: Housing and Community Development Acts, 1937-1990.

neighborhood rehabilitation and public facilities development; CDBG is also increasingly used as a component in financing affordable housing. The Urban Development Action Grant (UDAG) program adopted in 1977 to leverage private investment in distressed areas was terminated in 1989. Two earlier urban revitalization efforts—the Economic Opportunity program (1964-1971) and the Model Cities program (1966-1974)—were attempts to follow comprehensive improvement strategies in distressed areas, involving opportunities to improve the status of low-income residents as well as undertake physical and economic development. Federal legislation to support Empowerment Zones and Enterprise Communities through incentives under the federal tax code was adopted in 1993 (see Table 1.2).

The Mix of Programs, Assistance Mechanisms, and Delivery Systems

The mechanisms and systems by which HUD program assistance is delivered have also undergone change. Beginning with the adoption of the CDBG program in 1974, "block grants" rather than categorical assistance have been increasingly used; the adoption of the HOME housing assistance program in 1990 continued the "block grant" approach. Public

TABLE 1.2 The Cycle of Federally Assisted Urban Development Programs

Program	*Year Enacted*	*Terminated*	*Years in Operation*
Urban redevelopment or renewal	1949	1974	25
Economic opportunity	1964	1971	7
Model cities	1966	1974	8
Community development block grants	1974	—	22
Urban development action grants	1977	1989	12
Empowerment zones or enterprise communities	1993	—	3

SOURCE: Housing and Community Development Acts, 1949-1993.

housing modernization funds use a block grant formula based on documented needs. The allocation of funds by formula is also contained in the HOME program that uses a formula related to local housing need for rental assistance and housing rehabilitation, substantial rehabilitation and new construction. Significantly, the HOME program requires a matching contribution at the local level, reflecting the growing housing development pattern in localities of public-private partnerships as funding vehicles. Although the traditional public housing program and the Section 202 program for the Elderly or Handicapped are currently funded independently by project-based, up-front development grants, backed up by operating subsidies, this could change if the HUD proposal to establish a consolidated block grant program for housing development is adopted by Congress.

HUD's Social Service Functions

Another trend in HUD functional development is the increasing use of HUD funds for social services, largely through separate programs, including Housing Counseling, Community Partnerships Against Crime, the Youth Build Program, Early Childhood Development, and Neighborhood Development Grants. In its budget proposal for FY 1995, HUD proposed the consolidation of homeless assistance, including the absorption of the Food and Shelter program administered by the Federal Emergency Management Administration (FEMA); the FEMA transfer was rejected by

Congress. The Congregate Housing Services Program (CHSP), providing support services to elderly families in federally assisted housing to aid them in maintaining independent living, is funded on a matching share basis with 40% provided by HUD, 50% by local resources, and 10% by participating households. The Family Investment Centers, established in or near public housing developments to provide an integrated, accessible package of services to assist in family self-sufficiency, provide 100% HUD funding for the physical development of the Centers and 15% of the costs as a matching share of service costs. These last two programs, as well as some others, would be eliminated or consolidated under the proposed restructuring of HUD programs described above.

A Refined HUD Mission and Supporting Policies

Although the consolidation of HUD's programs and the streamlining of assistance and allocation mechanisms are essential for program management and flexibility, these changes do not address the need to integrate the HUD functions program under a more clearly defined HUD mission linked to a comprehensive national urban policy. The current array of HUD programs, or the proposed block grants, do not relate HUD's mission to the primary missions and resources of other federal departments and agencies. As noted above, the increasing use of HUD funding for income support (through Section 8 leases and operating subsidies for assisted housing developments) absorbs a major component of HUD's funding. Other critical issues concern the current statutory and administrative policy of focusing occupancy in assisted housing developments on the lowest-income families, rather than on mixed-income developments, and on a policy that neglects the opportunity to relate housing assistance to the HUD community development programs that could provide sound and livable neighborhoods as settings for assisted housing developments. Also, as noted above, the increasing use of HUD funding for social services subtracts from HUD's resources for its traditional mission of improving the supply and condition of housing and the physical and economic revitalization of neighborhoods and communities.

The mission set forth in the 1965 Act that created HUD had a two-pronged emphasis: to renew the declining areas of central cities and to guide the development of surrounding metropolitan areas. Over the more than 30 years of its existence, the Department has shifted its mission in response to the goals of different national Administrations and HUD

Secretaries, ranging from efforts to create an industrialized housing industry under Secretary Romney (1969-1973) to the goal of making HUD an "anti-poverty" agency under Secretary Kemp (1989-1992). The position of HUD Assistant Secretary for Metropolitan Planning and Development was eliminated in the mid-1970s.

The Struggle for a National Urban Policy

Efforts to create a national urban policy have been undertaken since 1970, when President Nixon's Council of Urban Affairs under the direction of Daniel Patrick Moynihan articulated 10 basic urban issues requiring attention; in that same year, the Congress established the requirement for a report on urban growth. Although the "growth" part of this report was eliminated by legislation in 1977, the requirement for a two-year urban policy report was established in 1978 and produced a series of reports by successive administrations. The reports of 1978 and 1980 were the most prescriptive, both in terms of defining urban needs and proposing federal actions to address them. However, these actions were never implemented and beginning in 1980, urban policy was not perceived or approached as a discrete issue but as a residual element of macroeconomic policy, returning responsibility to state and local levels and to personal decision making (Nenno, 1988). In addition, there was no established national housing policy but rather an unstated policy resulting from the interaction of market forces.

The absence of a comprehensive national policy responding to the needs of urban areas has left the Department of HUD without a basic framework within which to define and carry out its special mission and has contributed to the unplanned way in which HUD programs have evolved and shifted over the last 30 years. In 1995, several factors are generating pressures to renew the concept of a national urban policy, including the changes in the economic, physical, and social structure in urban areas, that is affecting the welfare of the nation as a whole and the growth of new programs and strategies at state and local levels, as described below. A major factor in stimulating new interest in the creation of a national urban policy, including a housing component, is the changing character of urban areas. Leading this change are dramatic shifts in the structure of the American economy away from manufacturing employment toward service industries, as well as toward new high technology jobs; these shifts alter the location as well as the substance of economic activity. This change has been accelerated by the growth of worldwide markets and the need to compete in an international arena. In the words of one urban scholar: "Fundamental policy issues surrounding the future of American cities are at stake . . . the polarization of urban areas and labor markets within them, the rise of an underclass, the deterioration of neighborhoods, infrastructure and services are closely related to economic change . . . the question is how to get federal policy makers to see the importance of cities as closely related to national economic and social policy" (Hanson, 1983, 1986). In addition, it is increasingly recognized that the traditional relationship between central cities and their metropolitan or regional areas is no longer valid. The central city is no longer the prime focus of urban life but is in an interdependent relationship involving new geographical and functional arrangements of industry and service activities (Blakely & O'Connor, 1988). The urban region is being recognized as the focus of urban life in the future, both within the nation and internationally (Peirce, 1993). These changes have commanding implications for the economic future of the nation as a whole and particularly for affordable housing and urban development. Housing initiatives must be increasingly linked with urban development efforts to create sound neighborhood and community life in new patterns of economic, physical, and social development related to the urban region under the framework of a national urban policy. Current trends to decentralize federal assistance under a concept of block grants without a national urban policy could work counter to the need to create a balanced and equitable development across the country.

■ Emerging Local and State Capacity

Localities and states are involved in housing and urban development activity in two ways: the use of general governmental powers relating to land use and zoning, building and housing codes, and real estate taxation; and the direct funding of housing and urban development activity. Increasingly, beginning in the mid-1980s, these two areas of action have been combined in more integrated approaches. The upsurge in local and state housing and urban development initiatives in the last decade has been widely attributed to the significant cutbacks in federal assistance. Between fiscal years 1981 and 1988, low-income housing assistance was cut by 75%, Community Development Block Grants by 32%, and Urban Development Action Grants by 100%. But the reasons for these initiatives relate not only to federal funding cutbacks but to the growing maturity of capability based on 40 years of public agency experience under federally assisted programs, new partnerships between public and private entities, and the new capacity of nonprofit community organizations. There was also increased sensitivity to the lack of affordable housing for lower-income households (particularly the elderly, the homeless, and the physically and developmentally disabled); the lack of homeownership opportunities for first-time home buyers; the extent of substandard housing and deteriorating neighborhoods; and the lack of housing at affordable prices to accommodate the workers related to business development and job initiatives (Nenno, 1989).

The Use of General Governmental Powers

Traditionally, states have not been prime actors in the use of general governmental powers for housing and urban development, but they have played very important roles in establishing the authority and standards for locally administered functions, such as building codes, zoning and land use regulations, minimum housing standard codes and real estate taxation. Two principal mechanisms used by municipalities since the 1920s to control land use are zoning and subdivision regulation. Another significant area of municipal influence are building codes that in some cities date back to early American history. But it was only after disastrous fires in Chicago (1871) and San Francisco (1906) that serious attention was given to local comprehensive building codes, and this involved slow progress since the model codes were first prepared by local building officials in 1915. Another area of action influencing housing and development was the evolutionary adoption of housing occupancy codes that governed health and safety conditions. A major expansion in zoning, building, and housing occupancy codes came after 1954 when the federal government required that cities and other public bodies adopt housing, zoning, and building codes as a condition of receiving federal urban renewal assistance; this requirement, known as the "workable program," resulted in the adoption of building codes in over 3,000 localities. Yet another municipal function affecting housing and development is real estate taxation; increasingly, local governments have learned to use the tax system to promote housing and guide development with tax abatements and other favorable incentives (Nenno & Brophy, 1982). A 1991 analysis of the use of land use functions to promote housing (housing linkage with private development, housing preservation and replacement ordinances, inclusionary zoning, and rent control) documented that in 133 cities with populations of over 100,000, from 12% to 19% used these mechanisms (Goetz, 1991).

States have also shown progress in establishing statewide codes. By 1966, four fifths of them had one of the following: statewide plumbing, electrical, boiler, or elevator codes. Almost half of them had a statewide fire code or regulations—but only five states had enacted statewide building construction codes (U.S. Advisory Commission on Intergovernmental Relations, 1966). By 1973, 15 states had statewide building codes, 28 had preemptive laws governing factory-built housing, and 38 had preemptive regulations for mobile home construction (McCollom, 1973).

Another area that helped shape housing and urban development was comprehensive community planning, stimulated by the federally assisted "701" planning program enacted in 1954, providing matching grants to state and local planning agencies; this program was terminated in 1981. All these general governmental functions continue to influence the course and direction of housing and community development.

The Growth of Direct Functions: The States

The federal government was the prime mover in stimulating direct functions to construct low-income housing, clear slums, and redevelop blighted areas. Beginning in the 1930s, it provided assistance to local public housing agencies to construct and manage low-income housing. A few states followed this example by creating state-assisted public housing programs. These early initiatives established the state and local governmental powers to undertake direct action to improve the supply and condition of housing, and to renew declining areas. The national government continued to lead this effort with the creation of the urban redevelopment program in 1949 and with additional mechanisms to finance low- and moderate-income housing. The experience gained under federal assistance programs stimulated interest in developing state and locally based programs, beginning in the 1960s and rapidly expanding in the 1980s with cutbacks in federal assistance.

A major surge in state activity in housing came with creation of State Housing Finance Agencies, beginning with New York State in 1960. By 1981, such agencies were established in 44 states and had assisted in the financing of 305,000 single family homes and 296,000 multifamily rental units (National Council of State Housing Finance Agencies, 1981). During the 1970s, the activities of these agencies were expanded through the use of tax-exempt mortgage bonds designed to lower the cost of housing for lower-income families; by the 1980s, these state agencies had assumed a variety of housing interventions designed to cushion the impact of declining economies on the housing industry (Massachusetts Housing Finance Agency, 1993). In addition, cabinet-level Departments of Housing or Development (or both) or divisions within other State Departments were taking on a variety of new functions. By 1980, 13 states had established housing rehabilitation grant or loan programs and 15 had provided tax incentives for housing (U.S. Advisory Commission on Intergovernmental Relations, 1985). Between 1980 and 1989, it is estimated that more than 300 new state housing programs were enacted. A 1992 report documented that 50 states administered more than 600 affordable housing programs, including homeownership assistance, rental housing development, special needs housing for elderly, handicapped, or homeless persons, as well as economic development and technical assistance (Petherick, 1992). At the same time, an increasing number of states were establishing state housing trust funds for low-income housing often with dedicated sources of revenue, such as real estate transfer taxes or other real estate-related sources; by the end of 1992, it was reported that 37 states and the District of Columbia had established housing trust funds (Center for Community Change, 1992, 1993). As of 1990, four states had also established statewide housing partnerships to assist in the development of local capacity to carry out housing and community development activity, including local political jurisdictions and community nonprofit housing organizations (National Association of Housing Partnerships, 1993). Also, 26 states had created "enterprise zones," providing tax incentives for businesses to locate in economically distressed areas. A few states, notably New Jersey and California, had undertaken "fair share" plans to pursue low-income housing goals and counterexclusionary housing practices.

■ Growth of Direct Functions: Localities

Traditionally, federal government assistance for low-income housing, urban development, and redevelopment were administered by separate authorities established under state enabling legislation, beginning with the creation of local housing authorities in the 1930s to administer federal public housing assistance and after 1949, by local redevelopment authorities or joint housing and redevelopment authorities to administer federal assistance for community development and redevelopment. Following the 1974 Housing and Community Development Act that created the Community Development Block Grant program, general local governments, as well as states, became the administrators of federal assistance.

Experience in administering federally assisted programs generated a new capacity in local government to carry out housing and community development activity. One evidence of the new capacity was the creation of local housing finance agencies to take advantage of public agency powers to issue tax-exempt bonds at below-market interest rates, making it possible to reduce housing costs for low-income families. Local agency issuances of tax-exempt mortgage revenue bonds increased from $1.5 billion in 1981 to $4.5 billion in 1985. Local housing authorities and redevelopment agencies also expanded their activities beyond those assisted by the federal government. These included initiatives such as low-income homeownership, acquisition, and rehabilitation of housing for homeless, handicapped, and mentally ill families. Increasingly, these agencies joined in partnership with private entities to finance joint housing development (National Association of Housing and Redevelopment Officials, 1993, 1994; Nenno & Colyer, 1988).

In addition, the experience under the federally assisted Economic Opportunity Program (1964-1971) and the Model Cities program (1966-1974), although short-lived, stimulated new actions by neighborhood-based nonprofit agencies to assume a place in housing and community development. A survey in 1992 noted that 27% of the neighborhood-based community development corporations (CDCs) were born of the activist spirit of the 1960s, products of the "War on Poverty" and the civil rights movement. Often, these groups began as community service or community action agencies and later moved into community economic development, typically relating to specific local problems rather than to national movements (Vidal, 1992). This new capacity was to expand dramatically during the 1980s in locally based initiatives.

New low-income housing (increasingly called "affordable housing") and local community development activity assumed a variety of forms and a new range of participants. Many of the new activities were single-focused individual projects developed in response to a particular community need. In 1989, the U.S. Conference of Mayors identified 127 single projects in local communities covering a range of activities, including homeownership, affordable rental housing, home improvements and rehabilitation, housing for senior citizens, transitional housing, single-room occupancy housing and homeless housing (U.S. Conference of Mayors, 1989). Of the most innovative nonprofit ventures, 26 were documented in 1988 including, sweat equity housing, shared equity financing, linked bank deposits, low-income housing trust funds, lease-purchase homeownership, limited equity housing cooperatives, mortgage or rental assistance pools, and housing partnerships (United Way of America and Community Information Exchange, 1988).

Beginning in the 1980s, local initiatives matured beyond single-venture projects to community-wide permanent structures and financing mechanisms that pulled together the skills and resources of local and public agencies, private enterprise and nonprofit development corporations. These mechanisms included public-private partnerships, low-income housing trust funds and innovative uses of local powers, such as inclusionary zoning and tax-increment financing. Local private corporations, including financial institutions, established low-interest rehabilitation loan funds and equity funding for housing and community improvement projects.

Typically, *local housing partnerships* involve the merging of public and private funds in cooperative agreements to undertake joint ventures for community improvement, often involving assistance to neighborhood community development corporations to carry out housing or neighborhood improvements. Unlike single-project partnerships, these agreements involve an ongoing "program of activities" on a community-wide base and extending over time. Typical housing development ventures involve the use of tax-exempt bond financing by local government and equity investments by private enterprise; also, federal low-income housing tax credits provide incentives for private investment, with the partnership staff putting together the financing package. Housing rehabilitation ventures often involve low-interest rehabilitation loans by a consortium of local lending institutions. Using a broad definition of partnership, some 73 local partnerships in large and medium-sized cities have been recently identified (National Association of Housing Partnerships, 1993).

Most of the *local housing trust funds* that developed in the 1980s and into the 1990s involve dedicated sources of revenue, often real estate transfer or documentary taxes, hotel or motel taxes, loan repayments from government programs, or initial capital from the sale of government property, proceeds from bond issuances, appropriations, or transfers from housing agency reserve funds. Chapter 3 of this book identifies 80 state, county, and local trust funds. The level of funding and the administrative mechanism to allocate these funds varies; often, there are independent administrative bodies with members drawn from local government, private enterprise, and nonprofit neighborhood groups. Funds are used for a range of activities, including the financing, construction, acquisition, or rehabilitation of housing for lower-income households (Center for Community Change, 1993).

A number of local initiatives involve linkages with private commercial or residential development, including *impact fees on downtown commercial development* to offset the impact of such development on the local housing market; *inclusionary zoning,* requiring that a portion of new private housing development be reserved for low-income families; and *tax increment financing* that requires that a portion of the increased tax revenues from urban redevelopment projects be set aside for low-income housing for a specified period of time.

The current status of local and state initiatives and the capacity in housing and urban development can be described as one of a wide variety of initiatives and a growing capacity of institutions, approaches, and funding mechanisms. However, except in a few cases, these initiatives have not been rigorously evaluated to identify those approaches that are most effective. There is some movement to coordinate these activities under partnership structures and development strategies. But there are wide differences among jurisdictions and many states and localities that have yet to take leadership in developing new directions away from basic dependence on federal programs.

Local Innovations in Low-Income Housing, Depressed Urban Areas, and Severely Distressed Public Housing

This book documents a range of promising local initiatives directed at the revitalization of depressed urban areas and severely distressed public housing developments, as well as the construction and rehabilitation of affordable housing. An examination of these programs indicates that there are a number of cross-cutting findings. These findings identify matters for priority attention, if this experience is to be expanded in a sustained effort to revitalize distressed areas, including some public housing developments, and to generate a larger commitment to low-income housing.

Adequate and Dependable Financial Resources

Almost all the cases in this book use federal funding resources, particularly CDBG, the HOME program, and Low-Income Housing Tax Credits. Revitalization of severely distressed public housing is heavily dependent on HUD funding under the Severely Distressed Project Fund, as the chapters on the Chicago and Boston public housing revitalization illustrate. The FY 1996 proposals by the administration to consolidate CDBG and HOME into larger block grants could influence how they will be used in local communities, as well as the level of funding available for the type of efforts noted in this book. Most likely, there will be additional pressure to raise funding resources at the local level. Although local housing trust funds, public-private partnerships, and foundations have grown to be important sources of local funding, there is serious question as to whether they can make up any substantial reductions in federal support; the level of funds generated by trust funds varies widely. In addition, although many local trust funds are supported by ongoing dedicated sources of revenue, most are not dedicated by ordinance and are subject to local political decisions, as the withdrawal of the dedicated funds in the San Diego Trust Fund illustrates (see Chapter 10). In the case of the Cleveland Housing Network, the City's Trust Fund is funded out of its CDBG program and subject to competing demands for funding to meet moderate- and middle-income as well as low-income housing needs. Krumholz notes in Chapter 3 that "long-term financial and political support for the Cleveland Housing Network and neighborhood community development corporations is by no means certain." It is also evident that the funds required to undertake the comprehensive renewal of large, seriously distressed urban areas is very large. Perhaps the best example of this fact is the experience cited in Chapter 8 on the Sandtown-Winchester area in Baltimore, where large commitments of federal, state, local, and foundation funds were generated and targeted on this one area. Goetz concludes that such funding is "extraordinary" and not likely to be repeated in other cities, let alone in Baltimore. All this experience indicates the current critical nature of funding resources to carry out local revitalization projects, particularly if federal assistance is reduced. A realistic assessment would be that if advanced decay in inner-city neighborhoods is to be addressed, it will require a federal government commitment similar to that of the Model Cities program (1966-1974), as well as significant state, local, and private investments. Beyond federal funding resources, the

Boston Partnership Report documents unusual and probably not replicable leeway in federal regulations, for example, the extension of Section 8 assistance to 25 years and the elevated fair market rent level.

Comprehensive Neighborhood Improvement

An additional finding in the experience of the local revitalization programs documented in this book is that individual project or narrowly focused improvement initiatives will not succeed except in relationship to the stability and livability of the surrounding neighborhood. This is documented in both cases of public housing revitalization efforts. In Chicago, despite the successful renewal of Lake Parc Place, there are concerns about the surrounding neighborhood. Also, the plans to extend the MINCS program to other locations has serious impediments to face, in particular the need to break up the heavy concentrations of low-income and minority families in inner-city areas. This is a large societal issue beyond the influence of public housing alone (see Chapter 6). In Boston, a substantial influence in the successful revitalization of the Commonwealth development was "due to its location in a relatively stable, economically diverse and racially and ethnically mixed community." This was in contrast to the location of most public housing in Boston where socio-economic conditions in the surrounding neighborhoods are depressed (see Chapter 5). In terms of large-scale renewal of distressed urban areas, the South Bronx initiative, reviewed in Chapter 4, is built around the concept of "focused, comprehensive and long-term neighborhood renewal." This is similar to the approach in Baltimore's Sandtown-Winchester project that is "a comprehensive program of community revitalization incorporating significant housing rehabilitation and new construction, enhanced health and human service delivery, open space improvements, community building and job creation" (see Chapter 8). Inner-city revitalization of this scope has not been attempted since the Model Cities program, noted under the section on financial resources above. Both the South Bronx and the Baltimore programs have generated success with this locally focused approach, but they require very substantial commitments of resources, including those of the federal government, that may be difficult to replicate.

Physical Development and Community Empowerment

Most of the experience in the urban revitalization efforts cited in this book also reveal another common issue relating to the resolution of conflict between "a physical development improvement" approach and "a community empowerment" approach. This is a long-standing concern dating back to the experience in the urban renewal program where the emphasis on "physical redevelopment" ran into serious opposition from local neighborhood groups; the Project Advisory Committees (PACs) under urban renewal were not a sufficient response to the demand for resident involvement. Since that time, there has been an impressive growth in the strength and capacity of resident advocacy organizations, including neighborhood-based community development corporations. These organizations are now taking active roles in the direction and implementation of community improvement programs. There is still evidence that this issue has not been completely resolved. In the case of the Sandtown-Winchester program in Baltimore (see Chapter 8), there were significant efforts from the beginning to involve neighborhood residents and organizations. Yet, Goetz observes that some residents and observers complain that physical development has run ahead of community rebuilding; these critics suggest that the process is in danger of losing its community focus. In response, the point is made that many of the problems relate to physical conditions requiring a pace of redevelopment to show quick results, and that community-driven processes, although laudable, are simply not on the scale required. This issue has not yet been resolved in Sandtown, although at this point, it does not threaten the continued progress of the transformation process in the neighborhood. The evolutionary experience in resident involvement in the BURP Program and the Housing Partnership Program in Boston, along with recommendations for strengthening this process, provide useful models (see Chapter 2). Another perspective on this issue is seen in the Cleveland Housing Network, where Krumholz observes a gap between the stated goals of the "neighborhood movement" and the observed accomplishments, including the goals of neighborhood empowerment of the poor or neighborhood self-reliance. A critical factor in the Network's success is the supporting network of public agencies, private funders, and corporate philanthropy which has resulted in neighborhood programs providing most of their benefits to low-income families. Attempts to impose the broader agenda of the neighborhood movement on this coalition of interests runs the danger of upsetting this support structure. Part of the answer to this issue relates to the continuing need to increase the capacity and expertise of neighborhood CDCs and resident participation support mechanisms. Here, the initiatives of the California Mutual Housing Association, described in Chapter 9, could be a model for a permanent, flexible response structure available to provide technical assistance to resident groups involved in a variety of affordable housing ventures.

Local Community Consensus and a Development Strategy

Another crosscutting issue observed in the reports of the housing and urban revitalization programs described in this book are what Krumholz describes as the generally ad hoc basis in which development initiatives are undertaken, without relationship to an overall neighborhood or community redevelopment strategy. The past decade has seen the rapid expansion of the number of entities—public, private, and nonprofit—in local housing and neighborhood renewal activity. Yet, in only a limited number of local communities is there a framework of a total community revitalization plan that can engage these actors in a joint endeavor based on an agreed on consensus of interests. A recent evaluation of the CDBG program documented that in only a few communities are there neighborhood or community-wide improvement strategies. More effective results would be achieved if they built on the best community practice: concentrated investment, housing linked to community redevelopment, economic development, social service support, and citizen participation in neighborhood planning efforts (Urban Institute, 1994).

An important consideration in future national urban policy and federal assistance is how the initiatives and lessons of local and state initiatives can be incorporated to use these growing capacities and resources. This is particularly true if federal policy and assistance is focused on assisting housing and development initiatives developed on an urban regional base by public-private-citizen organization partnerships.

Changes and Challenges for the Future

The current pattern of affordable housing and urban development activity in the United States is one built on incremental approaches over the years. There is significant capacity and commitment at federal, state, and local levels by public, private, and nonprofit entities, but there is an uneven impact across the country and for the most part, no coordination around central strategies or relationships to the dramatic economic, physical, and social changes at work in urban areas. Few states or localities have developed commonly accepted, cohesive development strategies. There is no national urban policy to focus and target federal leadership and HUD assistance to guide American urban development in a changing environment. In addition, as described above, some federal policies, such as prohibiting mixed-income occupancy in federally assisted housing or neglecting to build an important link between federally assisted housing and federally assisted community development, run counter to current needs. The challenge is to mold the current capacity and resources among all segments and all geographic levels into a more coherent and productive partnership. In 1996, political directions at the national level are running counter to a national and coordinated effort. But the economic, political, physical, and social forces at work in American urban life could force a reconsideration of current directions in the future.

REFERENCES

Blakely, E. J., & O'Connor, E. (1988). *Suburbia makes the central city: A new interpretation of city-suburb relationships.* (Working Paper No. 485). Berkeley: Institute of Urban and Regional Development, University of California.

Center for Community Change. (1992, April, July; 1993, January, April). *News from the Housing Trust Fund Project.* Washington, DC: Author.

Center for Community Change. (1993). A summary of revenue sources committed to existing housing trust funds. *Current Topics from the Housing Trust Fund Project.* Washington, DC: Author.

Goetz, E. G. (1991). Promoting low-income housing through innovations in land use regulations. *Journal of Urban Affairs, 13*(3), 337-351.

Hanson, R. (1983). *Rethinking urban policy in an advanced economy.* Washington, DC: National Academy Press.

Hanson, R. (1986). *Urbanization and development in the United States: The policy issues.* Washington, DC: National Association of Housing and Redevelopment Officials.

Massachusetts Housing Finance Agency. (1993). *1992 Annual report.* Boston: Author.

McCollom, J. C. (1973). *Building codes: A general assessment of their status and impact on residential building.* Washington, DC: U.S. Department of Housing and Urban Development.

National Academy of Public Administration. (1994). *Renewing HUD: A long-term agenda for effective performance.* Washington, DC: Author.

National Association of Housing and Redevelopment Officials. (1993, 1994). *Agency awards of merit in housing and community development.* Washington, DC: Author.

National Association of Housing Partnerships. (1993). *A catalog of local housing partnerships.* Boston: Author.

National Council of State Housing Finance Agencies. (1981). *Survey of state housing finance agencies.* Washington, DC: Author.

Nenno, M. K. (1988). Urban policy revisited: Issues resurface with a new urgency. *Journal of Planning Literature, 3*(3): 253-267.

Nenno, M. K. (1989). *Housing and community development: Maturing functions of state and local government.* Washington, DC: National Association of Housing and Urban Development.

Nenno, M. K., & Brophy, P. C. (1982). *Housing and local government.* Washington, DC: International City Management Association.

Nenno, M. K., & Colyer, G. C. (1988). *New money and new methods: A catalog of state and local initiatives in housing and community development.* Washington, DC: National Association of Housing and Redevelopment Officials.

Peirce, N. R. (with C. W. Johnson & J. S. Hall). (1993). *Citistates: How urban America can prosper in a competitive world.* Washington, DC: Seven Locks Press.

Petherick, G. D. (1992). *State HFA housing catalogs.* Washington, DC: National Council of State Housing Agencies.

Roukema, M. (1994, July 21). Statement on HR 3838, The Housing and Community Development Act of 1994. *The Congressional Record,* p. H 6020.

United Way of America and the Community Information Exchange. (1988). *Raising the roof: A sampler of community partnerships for affordable housing.* Alexandria, VA: United Way of America.

Urban Institute, The. (1994). *Federal funds, local choices: An evaluation of the community development block grant program.* Washington, DC: Author.

U.S. Advisory Commission on Intergovernmental Relations. (1966). *Building codes: A program for intergovernmental reform.* Washington, DC: Author.

U.S. Advisory Commission on Intergovernmental Relations. (1985). *The states and distressed communities: The final report.* Washington, DC: Author.

U.S. Conference of Mayors. (1989). *Partnerships for affordable housing: An annotated history of city programs.* Washington, DC: Author.

U.S. Department of Housing and Urban Development. (FY 1988-FY 1995). *Budget summaries.* Washington, DC: Author.

Vidal, A. C. (1992). *Rebuilding communities: A national study of urban community development corporations.* New York: New School for Social Research.

Part I

Developing Success Stories

SUSAN S. FAINSTEIN

The first part of this volume traces the rise of community development corporations (CDCs) out of the ashes of the federal public housing program. Although a few not-for-profit housing organizations like Phipps Houses in the South Bronx (chronicled in Chapter 4 by Lynda Simmons) have a long history of housing involvement, most are relatively recent. In many respects, they were born of desperation. The abandonment by the federal government of any large-scale commitment to providing housing benefits cut off the source of direct public sponsorship. At the same time, the various federal experiments with private sponsorship of low-income housing, beginning with the 221(d)(3) program and proceeding through the Sections 235 and 236 and Section 8 programs, stimulated activities by community organizations to take advantage of governmental subsidies in the absence of action by municipal governments.

The success stories of organizations in Boston, Cleveland, and New York disclose how ingenuity and commitment at the neighborhood level can create results in many respects superior to those produced by government-sponsored housing programs. Characteristics of the community-based nonprofit sector that seemingly distinguish it from governmental bodies include much greater flexibility, highly dedicated staff, an ability to learn from previous experience, and a willingness to provide supportive services. Although the cases differ in terms of whether the impetus for housing programs was top-down or bottom-up, the resulting programs all reflected an openness to community involvement in planning and management that also distinguishes them from public housing. Even in terms of the scale of production, the area in which CDCs are most frequently faulted, the organizations described here have excellent records. Moreover, they demonstrate that short-run failures to produce large quantities of housing quickly are more than made up for by the long-term benefits of careful planning and outreach efforts.

A reading of the three case studies shows a number of commonalities. In all three places specialized organizations carry out housing construction and management. Even when, as was the case in Cleveland, the CDCs traced their roots to advocacy organizations, the role of neighborhood advocate did not stem naturally from that of housing provider and was abandoned. Dependent as these organizations are on outside funding and given the extremely staff-intensive nature of the job facing them, they apparently cannot simply function in both capacities.

The cases reveal that the activities involved in the provision of housing for low-income people greatly exceed the task of simply building it. One of the major drawbacks of the CDC approach is the amount of time and energy that needs to be put into raising financing and the staggering complexity of the funding sources involved. Nevertheless, the organizations discussed here proved astute at this task. A comparison of the cases shows variation in the principal sources of funding, implying that no single investor is responsible for the growth in resources available to CDCs. All three levels of government, as well as philanthropic organizations and private investors, have played significant parts. Consequently, on the one hand, housing production is less dependent on the vagaries of a single funder, as was the case under the public housing program, where spending levels varied sharply from year to year. But, on the other, the seemingly multiple opportunities for raising money available to non-profits almost all ultimately depend on public programs, either of direct appropriations, as in Section 8, or of tax credits, whether for philanthropy or for participation in low-income housing syndication. As Rachel G. Bratt points out in Chapter 2, the impact of current budget cutting and of the efforts to create a flat tax system at all levels of government may, therefore, devastate CDCs just at the point where they have shown highly praiseworthy results.

In addition to the amount of effort they put into seeking financing for their projects, the CDC staffs all play ongoing roles in the continued functioning of their projects. Housing turns out to be much more than shelter; rather, it constitutes a bundle of services offered at the site where people live. CDC staffs provide technical assistance to tenants organizations, shareholders in low-income cooperatives, and low-income homeowners. They have developed ancillary social services ranging from daycare to job training. Some CDCs have spun off economic development components.

Success in maintaining stable projects depends on insuring a mix of tenantry. Part of the flexibility available to CDCs is that, unlike public housing managers, they need not restrict occupation to the neediest. They also have much greater freedom to evict problem tenants. At the same time, this wider leeway raises questions of equity. The temptation toward skimming the potential occupant pool is great, and biases in that direction increase as present residents gain control over the selection process. Norman Krumholz in Chapter 3 points to the growing preference for subsidizing owners, expressing concern that community-based housing may prove exclusionary. In general, all the experiences related here raise the question of who will take care of the most disadvantaged in the absence of targeted governmental programs.

Housing providers in all three cities learned from prior experience. Either under the same auspices or through the formation of new organizations, they moved toward a model that avoided financial pitfalls and provided long-term support for residents. They have found additional strength through participating in citywide or areawide coalitions. They can now tap into a much bigger pool of experienced not-for-profit housing planners and managers than was the case in their earlier days.

The extent to which other communities can model themselves on these examples is problematic. Leadership and experience are essential, and these attributes cannot be acquired but must be developed. The national organizations like the Local Initiatives Support Corporation (LISC) and the Enterprise Foundation play an essential role in cross-fertilization, allowing fledgling organizations to profit from the performance of others.

As the national volume of CDC-sponsored production rises, the stories of organizations like the Boston Housing Partnership II, the Cleveland Housing Network, and Phipps Houses in the Bronx become more than isolated anecdotes. At the same time, total reliance on networks of non-profits for low-income housing production raises some nagging issues. What happens in cities so lacking in an organizational base and motivated leadership that no initiatives are generated? And at the more general economic level, does this new "post-Fordist" system of flexible production depend on the exploitation of staff and the employment of construction and maintenance workers at poverty wages? In her discussion of the unique qualities of Phipps Houses, Lynda Simmons comments how unusual it is for such an entity to have in its employ professionals earning a decent wage with reasonable benefits. Not-for-profit housing is much more flexible than public housing, because it is not hindered by the bureaucratic requirements that afflict public authorities, and also because it is exempt from the desirable protections that public employment offers its workers.

On the whole, the cases of CDC success are heartening. They show the broadening of the mission of these groups from simple housing construction to education, economic development, and social service provision. They chronicle the development of coalitions among providers, moving them away from a focus on individual projects and toward a participation in coordinated planning and development of broad areas of their respective cities. Finally, they reveal that public-private partnerships need not only work to increase the gains of for-profit developers of office buildings and luxury housing. Instead, they can offer a vehicle for investment in poor neighborhoods that does not carry with it the threat of gentrification.

2

From BURP to BHP to Demo Dispo: Lessons From Affordable Multifamily Housing Rehabilitation Initiatives in Boston

RACHEL G. BRATT

Abstract

This chapter chronologically examines four major rehabilitation programs in Boston that were aimed at large-scale multifamily projects over a span of 25 years. The analysis focuses on the roles of residents and intermediaries, technical assistance, the choice of developers, minority and community participation, tenant relocation, subsidy levels, long-term affordability, and the availability of social services. Each succeeding program made modifications in its operations that acknowledged the experiences of the prior programs. This chapter identifies these lessons and assesses how they were incorporated. It also discusses the place of top-down public policy initiatives in community-based redevelopment.

■

During the 1960s, many urban neighborhoods lost significant proportions of residents as post-war growth carried increasing numbers of people up to the middle class and out to the suburbs. What was left behind was an aging housing stock—both multifamily and single family—and a poorer and less skilled population that was more likely to include people of color. At the same time, traditional urban entry-level jobs in manu-

AUTHOR'S NOTE: The author wishes to thank Robert L. Pyne (MHFA), Eleanor White (formerly at MHFA), Thomas Bledsoe (MBHP), and Langley Keyes (M.I.T.) for their assistance and insights.

facturing began to disappear as capital moved to nonunionized locales in the United States or overseas to reduce costs. Thus in many urban areas, there was a three-fold challenge: housing in need of significant upgrading and modernization, a population unable to afford the costs of these repairs, and a job market that provided few opportunities for low-skilled workers.

This chapter examines four major programs in Boston, spanning over 25 years, that were created to provide assistance to low-income households in need of decent, affordable housing. The Boston Rehabilitation Program of 1968 (BURP), the first and second programs sponsored by the Boston Housing Partnership of the 1980s (BHP I and II),[1] and the Massachusetts Housing Finance Agency's Demonstration Disposition program of 1994 (Demo Dispo) were all aimed at the large-scale rehabilitation of multifamily housing. Each succeeding program made explicit modifications in its operation that acknowledged the problems and experiences of prior programs and incorporated prevailing views about affordable housing. This chapter is presented as a linear progression of one program to the next, and it argues that significant learning took place as each new program was created. However, it is important to underscore that changes in program design also were a product of other experiences with affordable housing development that were occurring outside the framework of the four large-scale initiatives being presented here.

Also, despite the overriding argument that "social learning" was indeed occurring, the following analysis illuminates how knowing the right course of action does not necessarily result in following that path. Instead, awareness may be heightened and the optimum courses of action may be apparent, but insufficient resources, a lack of political support, or other constraints can prevent program designers or implementers from doing what they know best. Often, decisions are made with the conviction that "half a loaf is better than no loaf at all." Although partial programs may further serve to heighten sensitivity about the components that are lacking, the present level of understanding about inner-city multifamily housing rehabilitation programs suggests that it is likely unwise to settle for much less than a "full serving." The challenges of inner-city rehabilitation are now reasonably well understood and the ingredients of successful programs are relatively straightforward. The social learning of the past several decades has come at a price in terms of real dollars and personal costs. Future programs should strive to respect the progress that has been made and endeavor to maximize the benefits of the hard-learned lessons.

■ Background of the Initiatives

Until 1961, the public housing program was the only vehicle for providing federal assistance to nonelderly households in need of affordable housing. At that time, the first major subsidy program for moderate-income renters was enacted, the Section 221(d)(3) program.[2] But the program was modest in scope; only about 187,000 units were subsidized, of which 94% were newly constructed and the remainder were in rehabilitated buildings (U.S. Department of Housing and Urban Development, 1980, pp. 64-65). In 1969, Section 221(d)(3) was replaced by the significantly larger-scale Section 236 program.[3]

By the late 1960s, the period during which the first program in this study was launched, the housing situation in Boston, as in many other large U.S. cities, was serious and growing steadily worse. Between 1960 and 1970, there was a net loss of housing units, and the ability of many people to pay for decent housing was reduced by increases in costs. By 1970, deterioration and abandonment progressed unchecked, with some 1,300 abandoned residential buildings awaiting demolition, primarily in Boston's inner-city neighborhoods of the South End, Roxbury, and North Dorchester (Citizens' Housing and Planning Association, 1971).

According to a 1971 report by the Metropolitan Area Planning Council, low- and moderate-income households were expected to continue experiencing serious shortages of adequate housing. Black and Spanish-speaking groups were expected to face additional problems in locating available housing, due to discrimination. Citing an insufficient amount of new construction for those most in need, the report warned that "unless there is an increase in public action to remedy the situation, the number of deficient and substandard units is projected to remain virtually constant. . . . up to 1980" (Metropolitan Area Planning Council, 1971).

Another report released at about the same time that included information on housing conditions in the Boston Model City area (including much of Roxbury and sections of Dorchester) noted that "only 30% of the occupied units are classified as in 'good condition.' Almost half of the units need minor repairs, and one of every five units needs major repairs. The remainder (3%) are evaluated as being beyond repair" (Justin Gray Associates, 1970, p. 13).

The housing problems were, however, only part of the overall depressing urban landscape. The middle and late 1960s were a period of major social unrest, as cities all across the country (including Boston) experienced racial riots. At the federal level, several presidential commissions were formed and major reports, detailing the plight of the urban poor and minorities, were released.[4] In addition, two key pieces of federal legislation were enacted in 1968, the Civil Rights Act and the Housing and Urban Development Act. The latter created new programs, including the Section 236 program noted above, to expand affordable housing opportunities for low- and moderate-income people. Again, new housing dominated. Of the more than 450,000 units subsidized under this program, only about 10% were rehabilitation projects (U.S. Department of Housing and Urban Development, 1980, pp. 64-65).

Thus, as this story begins, the federal arsenal for dealing with the nation's existing stock of multifamily housing was negligible. Programs emphasized production of new housing and private landlords were expected to continue to provide affordable housing. But, as outlined earlier, the nonsubsidized affordable housing market had broken down in many locales by the late 1960s. Slowly, states and cities came to understand how they could use the federal programs to meet the shortage of affordable rental housing. Boston has been a leader, first in utilizing the federal programs and later in designing and implementing creative local initiatives.

This chapter chronologically describes each of Boston's four major multifamily housing rehabilitation programs. For each, the context in which the program was launched is presented, along with an outline of its goals, a description of how it was organized, major issues that unfolded in the course of the program, and an overview of the most important outcomes. More specifically, this chapter examines the ways in which the four programs approached a variety of concerns and challenges, including the role of residents, the role of intermediary agencies, technical assistance during the development process, the choice of developers, minority or community participation in development and construction, the relocation of existing tenants, the adequacy of the subsidies and of the rehabilitation, the adequacy of reserves, the commitment to long-term affordability of the housing, and the availability of social services. Although not all of these issues are relevant in all the programs, each one emerged as important in at least two of them.

For the three programs following BURP, the explicit lessons learned from the earlier initiative(s) will be highlighted, as well as an assessment of the extent to which these lessons were incorporated into the new program. To the extent that these programs are all top-down as opposed to bottom-up—originating outside the community rather than by the residents of the buildings included in the projects—they provide a good opportunity to assess the ways in which public policies can be launched and embraced at the local level. However, in articulating that all the programs were created by entities other than neighborhood residents, it is important to point out that there is a great deal of variation among the implementing agencies concerning how important they view the role of the community in carrying out the programs. The chapter concludes with observations about the role of top-down public policy initiatives in community-based development.

■ Boston Rehabilitation Program (BURP)

In late 1967, HUD's first Secretary, Robert Weaver, announced the allocation of $24.5 million in Section 221(d)(3) funds for the rehabilitation of some 2,074 units of multifamily housing located in the Roxbury-North Dorchester area, a predominantly African American community. With an emphasis on production and speed, the Federal Housing Administration's (FHA) processing time for project approval was to take no more than 60 days, and the actual rehabilitation was to be completed within six months by five all-white development teams. As the program unfolded, however, scale and speed emerged as far less important than who did the development, how residents were affected, and the quality of the construction.

During the ceremony marking, the formal announcement of the program, a local black activist interrupted the Secretary's remarks and voiced a series of complaints that foreshadowed much of what turned out to be problematic with BURP, as well as important lessons that were eventually learned about large-scale multifamily housing rehabilitation efforts:

> The program being dedicated has given no consideration to local developers, nonprofit developers, cooperative ownership or local management. It has been marred by racial discrimination in employment and inadequate relocation procedures.
>
> The FHA has shown that it can move with unprecedented speed—to give high profits to developers from outside the community and to establish a huge preserve for exploitation by absentee landlords.
>
> Apparently, Secretary Weaver's department can move quickly only when it is operating against community interest.[5]

According to M.I.T. Professor Langley Keyes (1970), "In 1967 Roxbury, like every other black community in America, was becoming more politically self-aware. Seasoned by the politics of the War on Poverty, hardened by urban renewal, the scene of two 'rebellions' in less than one year, the Boston black community was a complex maze of leaders, structures, and organizations" (p. 25). One such leader, Melvin King, recommended that steps be taken to "involve members of the community in meaningful ways . . . [including] training programs for potential black developers, contractors, and rehabilitation workers. . . ." (pp. 49-50). In response to the community's protests, six months after the initial announcement of BURP, two additional development teams comprised of African Americans were named to rehabilitate 200 units of housing, representing over $2 million in loans. In addition, a more confrontational process resulted in the hiring of some 300 black workers who participated in training programs and in the actual rehabilitation.

Community representatives also were irate about the lack of attention to the relocation of existing tenants, with an estimated 1,500 to 1,700 families slated to be displaced. According to Keyes (1970), although the original BURP model stated that displaced residents should have first consideration for the rehabilitated units, they were not guaranteed the right to return to their unit or move to another BURP unit. "By the time negotiations [with tenant and community activists] had ended, not only were BURP tenants given the right of first return, but it was agreed that no units were to be leased on the open market until all BURP needs had been met" (pp. 78-79).

Another flaw in the original BURP design that was addressed through a process of negotiation was the omission of any explicit role for tenants who were going to be affected by the rehabilitation. The relationship was adversarial, and tenants were quick to mobilize around poor conditions in the buildings, even going so far as calling rent strikes. Again, according to Keyes (1970), "The days of docile tenants, willing to tolerate whatever level of service the landlord offered, were over. BURP had introduced the local tenants' group as a permanent element in the relationship between landlord and tenants in Roxbury" (p. 91).

Further adding to BURP's problems was that it did not include adequate financial reserves to deal with unforeseen expenses. Although BURP had anticipated that tenant incomes would increase over time and that any additional costs would therefore be covered, this failed to happen and costs, particularly the price of fuel, skyrocketed in the early 1970s, contributing to the ultimate foreclosure of the BURP units.

An assessment of the outcome of BURP is, at best, mixed. On the one hand, many apartments were in better overall condition after the rehabilitation than before. On the other hand, the quality of the construction was "inferior to previous 221(d)(3) rehabilitation work done in Roxbury in similar buildings" (Keyes, 1970, p. 149). Also, summarizing the short- and long-range outcomes of BURP, Wendy Plotkin, a former graduate student, and I observed the following:

> Three of the five original developers did not finish their work within the six-month deadline, taking up to a year and a half. One developer did not finish at all, and HUD foreclosed during the construction period. . . .
>
> Problems continued after rehabilitation ended. Poor management and rising operating costs generated serious financial problems; by the mid-1970s, HUD had foreclosed on over half the units, representing the parcels of four of the five original developers and both minority developers. In 1982, HUD took possession of the remaining units. . . . In 1986, HUD initiated plans to dispose of the BURP buildings as part of a larger package of HUD-held properties in Boston, known as the Granite Properties. (Bratt & Plotkin, 1989, p. 295)

On a more conceptual level, Keyes evaluated BURP as follows:

> The history of the project demonstrated that speed and size are not enough to insure success . . . [defined as] rapid completion of the rehabilitation within the construction allowance granted in the original mortgage. For only when a project is structured to deal with relocation, local participation, and communication will that project avoid the expense and delay of confrontation, and thus fulfill its stated objective. To build in these components as an afterthought or in response to pressure is clearly less efficient than to include them at the beginning. . . .
>
> The BURP experience makes clear the need for local participation in any project which affects the lives of residents of central-city neighborhoods. Given the tenor of the times, it is no longer sufficient for public and private decision makers to be concerned solely with a product—such as rehabilitated housing. How that housing gets produced has become as important an issue as the fact that it is produced at all. (Keyes, 1970, pp. 156-157, 166)

Thus, the overriding lesson of BURP is that the process is at least as important as the product. Involvement by the local community turned out to be essential. Those closest to the program came to realize that at least some of the development entities and workers must have a stake in the community and that affected residents must be brought into the process. These lessons were not lost on the programs that followed.

Boston Housing Partnership— The Demonstration Program (BHP I)

In the early 1980s, at the start of the Reagan era, many cities across the country realized that federal resources for housing would be significantly diminished, and that local initiatives in the form of public-private partnerships were needed. In 1983, Boston responded with its own such effort. The Boston Housing Partnership was created jointly by the city's development agency with significant support from Mayor Kevin White, and several large banks (notably the chairman of State Street Bank and Trust Co., William Edgerly), as well as neighborhood organizations, with the goal of converting deteriorating and abandoned housing into decent, affordable dwellings for low- and moderate-income families. Begun in 1984, the BHP's first project, known as the Demonstration Program (or BHP I), entailed the rehabilitation of 701 units of housing.

From the outset, BHP I was different from BURP. First, whereas in BURP, the lead agency was a federal department (HUD), in BHP I, the organization in charge was an entity spawned from City Hall and the local business community. In its effort to develop the Demonstration Program, BHP I was committed to solving local problems. The Boston Housing Partnership was joined by the Massachusetts Housing Finance Agency (MHFA), a highly sophisticated agency that has consistently been in the vanguard of providing funding for housing development,[6] and the Bank of Boston that underwrote the mortgage financing. As part of its mission to finance both multifamily and homeownership units, MHFA has promoted the provision of affordable housing, and it has been a major innovator in providing services to residents of its developments. In the BHP I project, MHFA originated the mortgages that were then sold to the Federal National Mortgage Association (or Fannie Mae), with MHFA retaining the servicing of the loans.

Although BHP I was certainly better off than BURP in terms of having far more responsive and "in touch" oversight agencies, the loss of HUD as a key player in BHP I[7] and the overall federal retrenchment in housing that took place in the 1980s meant that the financing package was significantly more complex. Whereas both the subsidy and the financing had been provided to the BURP developers directly by HUD, BHP I represented a veritable patchwork of financing, the norm for affordable housing development since the early 1980s:

> In the BHP Demonstration Program, multiple financing sources were a necessity, attributable to the lack of direct and deep subsidies. The resulting complexity of the financial package had a snowballing effect on the problems it caused. Compounding this, the scope of the program, the buildings targeted and the decision to use nonprofit community-based sponsors contributed to a sense of caution among many key actors, most notably financial ones. Devising comfortable ways of participating meant trying to avoid arrangements perceived as too risky. Not surprisingly, the greater the effort to spread or share the risk among several parties, the more complex the financing package. Several plans were proposed and turned down and the BHP initiated the Demonstration Program with an incomplete financing arrangement. . . . The Demonstration Program, like its predecessors, was forced to underestimate costs. (Bratt & Plotkin, 1989, pp. 305-306, 308)

In all, BHP I involved $22.3 million of first mortgage financing. In addition to the city's $4.5 million CDBG grant, the deal also included $10.7 million in equity financing, $430,000 of private foundation grants, and $3,414,000 in annual subsidy payments from a combination of the federal Section 8 program, the state's rental assistance program (known as MRVP) and another state-funded subsidy for affordable multifamily housing (known as the SHARP program) (Metropolitan Boston Housing Partnership, n.d.[a]).

A second difference between the two programs was the way in which BHP I made an effort to correct one of the defects of BURP: inadequate cushions to deal with increases in operating costs. Despite the effort to create adequate reserves in BHP I, less than one year after construction was completed, all but $400,000 of the $5 million that had been set aside was gone. As a result, BHP I was launched with shaky finances, and its fate mirrored the problems encountered by BURP.

A third key difference between BURP and BHP I was that the latter enjoyed the considerable support of its sponsoring agency, the Boston Housing Partnership. Due to its continuing monitoring and support functions, the financial difficulties encountered by the nonprofit sponsors of the BHP I developments were able to be addressed. BURP, on the other hand, did not use an intermediary model and did not include a long-term support system that would help to ensure the overall success of the developments.

A final way in which BHP I and BURP differed was in the decision about who should develop the units. Unlike BURP's commitment to select the most experienced for-profit developers without any consideration of community linkages, the BHP I developers were all nonprofit, community-based sponsors. Ten such groups were selected to participate in the program, some with a considerable track record in developing affordable housing, while others were completely inexperienced; at least one group was created specifically to participate in the new program.

Nevertheless, the resources flowing into BHP I and the jobs that were created did not all benefit the local community. Many of the contractors were neither minority- nor community-based, and several downtown banks, law firms, and consultants profited from doing business with BHP I. Despite the uproar about the need to hire local workers to participate in BURP's rehabilitation, BHP I did not initially target minority and local residents as recipients of the jobs that were to be created. Two months after the inception of the program, however, a memo sent by the BHP to the newly chosen sponsors urged them "to make every effort to involve minorities and minority-owned businesses in their projects" (Bratt & Plotkin, 1989, p. 297).

> In the end, the BHP performed quite well in terms of local and minority hiring, primarily at the subcontractor level, with Boston residents and minorities working a majority of hours on the job. . . . Unfortunately the CDCs [nonprofit community development corporations] had difficulty locating general contractors who could meet the performance bond requirements; only two of the nine general contractors chosen were minority. The tenth community-based group . . . is minority controlled and served as its own general contractor. (Bratt & Plotkin, 1989, p. 297)

According to Thomas Bledsoe, Executive Director of the Metropolitan Boston Housing Partnership (MBHP), although specific minority business enterprise (MBE) hiring goals were not set for BHP I, the actual performance was reasonably good:

> The overall MBE utilization on the development side was 43% of total construction costs (soft costs not available). Forty-nine percent of the total construction work hours went to minority workers.[8]

Although it may be somewhat perplexing why BHP I did not initially target jobs to the local community, by the 1980s, there were explicit city and state government requirements concerning local hiring on public projects. In the minds of the program designers, therefore, it may have originally seemed redundant to reiterate the importance of giving preference to local workers.

Although BURP did not originally take relocation into account, BHP I asked each sponsor to develop a relocation plan. Similar to the issue of targeting jobs to the local community, relocation had been so inflammatory in earlier periods that there was no way that it could be ignored by the time BHP came into being. On this issue, the designers of the program were explicit at the outset about the need for closely examining the relocation process. BHP I required each sponsor to detail the relocation needs of the project as part of its application and to outline the extent and anticipated costs involved.

In many important respects, BHP I represented a marked improvement over BURP. Although residents were not central to the program, nonprofit community-based organizations were the developers, as opposed to the for-profit developers from outside the community that were used in BURP. Also, the overseeing agencies involved in BHP I, the Boston Housing Partnership and the MHFA (acting as mortgage servicer for Fannie Mae), provided a great deal more supervision and support to the developers than had HUD under BURP. Furthermore, to provide the nonprofits with the technical assistance they needed, BHP I allocated funds for this purpose, with the choice of who and what kinds of technical support were needed to be left up to the nonprofit developers.

Even though the quality and degree of rehabilitation had been a serious issue in BURP, this problem again surfaced in BHP I. Although final rehabilitation costs rose some 45% from the original budget, the net result was that the construction was still not adequate. BHP I was severely undercapitalized from the outset. In 1990, to provide an infusion of capital into those developments that had not received sufficient rehabilitation and were experiencing vacancies and high operating costs, BHP created a Stabilization Plan that involved aggregating over $1 million. According to the MBHP:

> In spite of these efforts, by 1992 nearly all of the ten developments were experiencing financial difficulties arising from rent caps instituted under the MRVP subsidy program; higher than anticipated operating costs and the effects on vacancy of having to comply with stringent lead paint removal laws. In addition, the economy was not nearly as robust as it had been in 1985 when the first apartments were renovated. With higher

> vacancies everywhere in the state, residents had more options about where to live and vacancies rose in the portfolio. The result was that initially 2 of the developments and then 6 more went into mortgage default during 1992 and 1993. (Metropolitan Boston Housing Partnership, n.d.[a])

During the fall of 1994, Fannie Mae, the holder of the first mortgages, committed $6.6 million as part of a complete financial restructuring that would cover a series of costs, including zero interest second mortgages to pay off the first mortgages, funds for capital improvements, and funds to cover delinquent accounts payable. As part of the new financial plan, some $275 per unit and $5000 per project were to be set aside to fund a new replacement reserve account (Metropolitan Boston Housing Partnership, 1994). As of late winter, 1995, it was still too early to assess the extent to which the new financial arrangements will allow BHP I to become economically viable.

■ Boston Housing Partnership II— The Granite Properties (BHP II)

Over half of the BURP properties and an additional 800 units, all of which had been rehabilitated by one developer, Maurice Simon, were packaged into a large-scale rehabilitation initiative in the late 1980s, known as the Granite Properties.[9] As early as the mid-1970s, only a few years after their initial rehabilitation, half of these units were in default and the physical condition of the buildings was poor. In 1982, HUD, in its role as mortgage insurer, assumed control of the properties as "mortgagee in possession." However, it was not until late 1988 that rehabilitation actually began under the auspices of BHP II. HUD's role in the interim was motivated by three broad sets of interests:

> To protect the federal financial interest in the properties, to minimize further outlays of funds and to minimize the federal government's long-term obligation to the properties by transferring ownership to the private sector wherever feasible. In fact, HUD's third concern was thought to be a means to the other two; it was believed that privatization would cut costs, or both. (Bratt & Morris, 1989, p. 149)

Between 1982 and 1988, city and state government agencies, nonprofit organizations, and tenants played active roles negotiating with HUD. The major issues revolved around whether HUD would provide adequate subsidies to do the necessary rehabilitation, and whether it would support the sale of the properties to nonprofit community development corporations, as opposed to private for-profit developers. Ultimately, HUD agreed to sell the units to one minority community-based developer and to a group of nonprofits, again brought together by the Boston Housing Partnership, and to commit significant subsidy dollars.

Following the pattern set by BHP I, BHP II also relied on a select group of nonprofit organizations for the development and ownership of the properties. By the mid-1980s, nonprofits in Boston were viewed, more or less unconditionally, as the developers of choice for affordable housing development. The problems encountered by the nonprofits that participated in BHP I, which were largely the outcome of the inadequate financing and rehabilitation, were in no way seen as a deterrent to continuing and expanding their role in future initiatives. In fact, several of the groups that had participated in BHP I were also involved in BHP II. The flaws with BHP I and its disappointing outcome were viewed as the result of structural defects with the program; blame was not placed on the nonprofits themselves.

BHP II also continued the commitment that minority contractors and workers play a key role in the actual rehabilitation, which had first surfaced as an issue in the BURP program, and that was supported in BHP I. Again, according to Thomas Bledsoe,

> BHP II achieved a 44% MBE rate for construction costs and 43% for soft costs. Fifty-eight percent of total construction hours were performed by minority workers.[10]

Drawing directly on the negative experiences of BHP I, as well as other preceding undercapitalized rehabilitation projects, the program designers of BHP II were committed to doing adequate rehabilitation at the outset and making sure that operating subsidies would be sufficient. BHP II relied on an elevated "fair market rental" as the major source of subsidy. (A fair market rental is the price of a housing unit, designated by HUD, that is used in calculating the amount of Section 8 subsidies. They are calculated for each metropolitan area and vary by bedroom size.) Most of the costs of rehabilitation were covered by the mortgage that, in turn, was amortized by federal Section 8 subsidies. BHP II program designers were able to convince HUD to allow rental levels and, therefore, the Section 8 subsidies to be based on a fair market rental set at 144% of the standard fair market rental for the Boston area. This was the key vehicle for maximizing the possibility that both the short- and long-range financial needs of the development would be met. In addition to HUD's commitment of Section 8 funds, some $18 million in additional financing was provided by investors purchasing Low-Income Housing Tax Credits through a syndication process, as well as from several city and state programs for affordable housing (Metropolitan Boston Housing Partnership, n.d.[b]).

By the time BHP II was launched, the housing advocacy community, public officials, and various nonprofit organizations, as well as the staff at MHFA, began to articulate the importance of combining a social agenda—a concern for tenant and community needs, not just the "bricks and mortar" of developing and managing the buildings—with affordable housing development. As for the MHFA, it began to outline a new mission for itself. During its first two decades, it had largely operated as a public purpose lender, providing funds to housing with some set-asides for low- and moderate-income households. However, by the mid-1980s, MHFA began to get directly involved in the social issues facing the residents in the housing developments that it financed. MHFA's adoption of a social agenda was part of a larger movement into this area among other Boston-area public and private organizations, including the BHP.

Several important initiatives were built into BHP II from the outset, rather than becoming available only after the program had gotten under way, as had been the case with BHP I. Some of the most important elements that were reflected in BHP II included the creation of the Tenant Assistance Program (TAP), the Inner-City Task Force, and the Resident Resource Initiative; funding for security personnel; and a Housing Court that was more amenable to enforcing eviction orders. The basic assumption underlying all these efforts was that the viability of the housing was inextricably related to the social conditions in the buildings, and that addressing resident needs was an essential ingredient of providing high-quality affordable housing.

The TAP program was created in 1984 to provide training both to property managers and tenants in the combatting of substance abuse. The goal was to deal with the "disproportionate amount of disruptive behavior and property damage that could be attributed to alcohol and other substance abuse in MHFA-financed multifamily housing" (Siflinger, 1994, p. 4).

The Inner-City Task Force was created in 1986 at the request of residents to provide a forum for residents, property managers, social service providers, the police, developers, and public officials to contend with issues concerning drug and gang-related violence, as well as overall economic conditions in the inner city. This effort was "working to reclaim affordable developments from drug dealers, vandalism and neglect, and to develop innovative programs to improve the quality of life for residents of urban subsidized housing" (MHFA, 1993, p. 21). MHFA's Youth RAP program was an outgrowth of the Inner-City Task Force, providing education, recreation, and employment opportunities to more than 3,000 young people living in MHFA developments. Despite the involvement of residents in the Task Force, residents did not have major roles in the development or implementation of BHP II.

The Resident Resource Initiative was formed in 1987 with support from several Boston-based foundations under the leadership of the BHP. This program has funded a portion of the salaries of ten Resident Resource Specialists who have been hired by ten local community development corporations. These individuals are "the cornerstone of the Initiative, providing a linkage between the CDC and the tenants, promoting organization and collective action among tenants, and providing residents with information and referral to social services and other resources" (Conservation Company, 1991, p. 1). According to the MBHP,

> residents are becoming full partners in the efforts to revitalize and empower the neighborhoods where they live. Residents are making their buildings safer, working actively with property managers, the police, and security guards to rid their buildings of drug dealers and gangs. They are participating in decisions about how their buildings are managed, spearheading efforts to plant trees and community gardens, and turning abandoned lots formerly controlled by gangs into playgrounds for children. (Metropolitan Boston Housing Partnership, n.d.[c])

Another program that was built into BHP II almost from its inception was the security program. Similar to the creation of the Youth RAP program, which was a response to tenant concerns, the security initiative also grew out of the Inner-City Task Force. With funding provided by MHFA, several local security companies provide on-site security personnel to MHFA-financed developments operating in high-crime and high-drug areas of the city.

Also, BHP II, developing during the era of the "expiring use crisis,"[11] was more sensitive than its predecessors to the need for the units included in the program to remain affordable over the long term. Hence, the nonprofit sponsors were given the option to buy out the investors (for-profit owners of the partnership) after 15 years, at favorable prices.

Finally, by the time BHP II was launched, the Housing Court had become more sympathetic to evicting tenants who were clearly in violation of the terms of their leases for such offenses as drug dealing or nonpayment of rent. According to Eleanor White, former Deputy Director of MHFA,

> In 1986 or 1987 we had several meetings with Housing Court Judge Daher. We explained that we only bring people into court for eviction proceedings when there is a clear cause and that we are not acting improperly against the tenant. Prior to that, when we brought tenants to court, Greater Boston Legal Services lawyers often ended up defending high-income criminals. We also brought them into the process and now when we bring someone in for an eviction we get very little opposition. Being able to evict disruptive tenants is absolutely essential to maintaining a safe and secure living environment for the other tenants in the building. We probably ended up evicting about 10% of the residents from the BHP II buildings.[12]

The lesson that security is an essential prerequisite for managing affordable housing has been a crucial one to learn. BHP II was the first large-scale rehabilitation initiative in Boston that fully embraced this idea and created a program to address the problem. The other key lessons that BHP II incorporated were the importance of working directly with residents (as well as with community-based nonprofit development corporations), the central role of social services, the need to safeguard the units as affordable over the long term, and the absolute necessity for rehabilitation and operating budgets to adequately meet the physical needs of the buildings.

It is too early to provide a final evaluation of BHP II. However, it is generally considered a successful program, with both the physical and financial needs of the developments remaining sound, and with tenants being provided a decent quality of affordable housing. As discussed below, the strengths of BHP II were acknowledged by the U.S. Congress.

■ The Demonstration Disposition Program (Demo Dispo)

The Demo Dispo program of 1994 represents the culmination of what has been learned from the three prior large-scale multifamily rehabilitation initiatives, as well as scores of smaller projects, that have been undertaken in Boston since the late 1960s. Although the BHP II program was not yet completed, Congress in 1987 authorized a new Demonstration program that was modeled after that initiative. The key aspect of BHP II that was incorporated into Congress's Demo Dispo program was a central role for strong local implementing organizations in the form of state housing finance agencies. Because BHP II had served as the model for the federal program, it is not surprising that Boston's Demo Dispo program is further along than the three other participating locales.[13] Twelve packages of HUD-owned properties, representing some 1,900 units in 140 separate buildings in the South End, Roxbury, and Dorchester comprise the Boston program. Taken as a whole, Demo Dispo emphasizes a central role for residents, economic development for the local community, and the provision of social services, while tackling the technical details and financial requirements of housing rehabilitation.

Probably the most important feature of the Demo Dispo program, which has slowly been embraced as a result of the experiences with prior rehabilitation programs, is that the residents must be at the center of the rehabilitation process. From the early days of BURP, when the lesson that those most affected must be included in the program, to the growing awareness of the need for resident participation during BHP I and II, Demo Dispo squarely places residents of the buildings slated for rehabilitation in control.

According to the MHFA's former Deputy Director, Eleanor White,

> The tenants are in control of the process. We tell them that we'll provide the tools, in terms of technical assistance and financial resources, but that they are responsible for getting the work done. This will entail their deciding the type of ownership structure that they prefer and hiring architects, contractors, property managers, etc.[14]

Based on MHFA's guidelines, there are three preferred resident-centered development models:

1. Negotiated sale to residents who want to move immediately to a 100% resident cooperative or other model of resident-controlled ownership of rental housing;
2. Negotiated sale to residents and a developer (either nonprofit or limited dividend) in a joint venture in which tenants have more than a 50% interest in the partnership; and

3. Sale to a limited dividend or nonprofit developer who is selected by a resident association and ratified by MHFA, following an open, competitive process. In this instance, the developer would pledge to train and support residents to move toward full resident control as quickly as possible, including the establishment of target date milestones for this process (MHFA, n.d., circa 1994).

Along with the commitment to include residents as key players in the program, the program appears to include the resources necessary to enable tenants to assume responsibility. Most significantly, the MHFA has earmarked $400,000 to $500,000 per resident organization for them to acquire the needed technical assistance, including help in packaging their ownership proposals in preparation for purchasing the properties. The MBHP is also providing a considerable amount of technical assistance to the fledgling tenant associations in their efforts to organize themselves and to take the necessary steps in the rehabilitation process. Providing both further funding for technical assistance, as well as direct support in the rehabilitation process, is the Community Economic Development Assistance Corporation (CEDAC), a quasi-public state-funded agency that assists nonprofits in their affordable housing and economic development initiatives.

Clearly, the success of transferring control and ownership to residents will be dependent on how effectively residents organize themselves and embrace the program. For some groups, however, this may be a difficult obstacle. Originally, the program called for all the deals to be closed by June 1996. However, due to changes in the national political climate, in which committed housing subsidies were rescinded by Congress, tenants were required to develop a draft disposition plan by August 31, 1995. Even operating under the longer time frame, some tenant groups were feeling pressed for time. According to Thomas Bledsoe,

> Many of the resident groups have been meeting four nights a week to hire their development teams or architects, attorneys and the development consultants. The resident groups have pushed MHFA very hard to slow down the process but have been largely unsuccessful due to MHFA's strong (and valid) concern about closing all the deals [on time]. Without this "political" constraint, more time would certainly be allowed for the residents to work through these critical and difficult decisions. The jury is still out on what impact the rushed schedule will have on the overall success.[15]

Although it will be quite a while before the final chapter on this critical phase of Demo Dispo can be written, prior research on tenant management corporations suggests that this type of top-down process may be problematic. The most successful resident management corporations in public housing having arisen out of a struggle between tenants and the housing authority, not from the efforts of outside organizers (Peterman, 1993). Whether Demo Dispo will be able to stimulate viable tenant groups in the absence of such conflict remains to be seen.

In addition to the important role for residents in Demo Dispo, another key feature is the goal that 80% of the aggregate controllable costs (excluding such "uncontrollable costs" as taxes and utilities, for example) be targeted to minority business enterprises. With some $100 million in funding earmarked for the program, this infusion of resources to minority firms is likely to produce the proverbial "shot in the arm" for scores of businesses. Again, Demo Dispo is incorporating the hard-learned lessons of its predecessors, particularly BURP, that included no goals for hiring minority businesses and initially excluded minority development teams.[16] Even BHP I and II, in which nonprofit organizations played key roles, minority participation was not an explicit goal. However, as noted earlier, there was substantial minority involvement in the construction process in both of these programs.

Another key difference between Demo Dispo and its predecessors is that this most recent program appears to be adequately funded, although it is far too early to say this with complete certainty. HUD is slated to provide the funds to cover the capital costs of the rehabilitation by drawing from the FHA insurance fund. It is anticipated that approximately $90,000 per unit will be needed to cover the costs of the needed repairs and rehabilitation. On completion of the construction work, the units will receive 15-year Section 8 project-based certificates or vouchers that only need to cover operating costs, because the capital costs will have been paid for up front.[17] Given the front-end capital being provided by HUD in Demo Dispo, the FMR will not be nearly as high as in the BHP II program.

Demo Dispo has acknowledged the importance of trying to build long-term affordability into the program, although it does not solve this problem. The initial Section 8 subsidies are committed for only 15 years. Although early discussions between HUD and MHFA indicated that two five-year renewals of the Section 8 subsidies might be possible, there is no guarantee that this will occur. Thus, there is the potential that this program will experience problems similar to those encountered in prior federal housing subsidy programs that have been facing expiring use restrictions. However, it is hoped that long-term affordability will be enhanced, because tenants will be in control of the buildings, and as a result, there will be an identity of interest between owners and residents, unlike the situation in the expiring use properties.

Finally, similar to BHP II, which incorporated a number of provisions into the program at the outset (e.g., the TAP program, the Inner-City Task Force, the Resident Resource Initiative, funding for security personnel, and a Housing Court that is cooperative in dealing with evictions), Demo Dispo also is benefiting from these significant initiatives.

There is a considerable amount of local enthusiasm about the prospects for Demo Dispo. Again quoting Eleanor White,

> We really feel like we have all the pieces in place. In order to have a positive effect on an inner-city community, several things are critical, particularly a resident-centered process and an agency that is both committed to and knows what it is doing. We have a strong belief in what we are doing. We have the luxury of "staying power"—we've been here a long time and we are able to fully devote ourselves to the difficult problems of inner-city housing.[18]

Furthermore, a *Boston Globe* editorial called Demo Dispo "a decisive and innovative step" that "will likely translate into many rightful homecomings" (*Boston Globe,* 1994). If the designers of Demo Dispo are correct, we may at last have an inner-city housing rehabilitation program that closely approximates the ideal. Evaluations being conducted by a team of researchers at the University of Massachusetts in Boston will hopefully provide a full picture of the strengths and weaknesses of Demo Dispo, as it continues to unfold. At the very least, it is certain that this most recent program has incorporated many important lessons from the past.

Conclusions

This chapter has explored the evolution of a series of large-scale rehabilitation programs in Boston. It is clear that a great deal of social learning has taken place. For the most part, later programs learned from earlier ones and, although the same mistakes were sometimes made, there has been a consistent forward movement with significant improvements in each succeeding program. Demo Dispo appears to have addressed most of the negative experiences of the past in what seems to be a remarkable conceptualization of the role of housing rehabilitation in the inner city. Going way beyond bricks and mortar, while still being highly attentive to what is needed to get the job done, Demo Dispo sees the rehabilitation effort as a way to engage residents, to create high-quality living environments, and to promote economic development.

Specifically, this chapter has witnessed the evolution of a "state-of-the-art" understanding of how inner-city multifamily housing rehabilitation programs should be designed and implemented. Many of the following lessons were neither recognized during the BURP program nor was their importance fully appreciated. Over time and through experimentation, we now have a comprehensive list of what appears to be the key components of a multifamily housing rehabilitation program:

1. an adequate financing package put together at the front-end;
2. a strong intermediary or sponsoring organization;
3. a central role for residents and/or community-based nonprofit organizations;
4. adequate security in the buildings;
5. social services for tenants;
6. involvement of the local (e.g., minority) development community;
7. adequate funding to allow for good long-term management; and
8. building long-term affordability into the program.

For all but the last concern, in which long-term affordability is only guaranteed for 15 years, it seems that the most recent housing rehabilitation program in Boston, Demo Dispo, has gotten it right—that the lessons of the past have been embraced by the new program. Although it will be years before a full assessment of Demo Dispo will be possible, BHP II, which was the immediate forerunner of Demo Dispo and has incorporated many of its features, is doing well. It is with a sense of both realism and optimism that one can predict an even stronger showing for Demo Dispo. At least for the present, our best wisdom about large-scale inner-city housing rehabilitation appears to be represented by this program.

Beyond the lessons about how to do rehabilitation, however, is the question about how government can best implement programs that are administered at the local level. In other words, how can a top-down process be embraced as a bottom-up initiative?

Demo Dispo suggests several guidelines for how bottom-up initiatives can be supported and nurtured through government programs. For any given program, however, these broad suggestions would have to be given concrete definitions to operationalize many of the phrases used below, such as "clear goals."

1. *The government program should have clear goals that are as consistent with community or low-income residents' needs as possible and create central roles for their participation.* Given what we have learned about government intervention in poor, urban areas it is not conceptually difficult to design programs that reinforce rather than conflict with a community's agenda. A key way to achieve this is by designing programs that are to be implemented through local nonprofit organizations and resident associations. Ownership of the project, in both financial and physical terms, should (whenever possible) be in the hands of not-for-profit entities and the government bodies in charge, and their designated intermediaries must be convinced that this is indeed the priority. This recommendation is critical, and it is made with a full appreciation of how difficult it is for any outside entity to stimulate a bottom-up movement, as discussed earlier.

2. *The program should make sure that timetables are sufficiently generous to allow time for the residents and community leaders to fully understand and embrace the program and to modify it to better meet local needs.* Often, involving the community can be time-consuming. It is important that the program be flexible enough to both absorb delays necessitated by a community-oriented process, as well as to make changes that arise during the period in which residents become engaged, and to begin to take ownership of the project.

3. *The government agency selected to represent the top-down interests should be skilled in its work, have a significant track record, and have a good working relationship with the local community.* Whether the initiative originates at the federal, state, or local level, one or more carefully selected state or local organizations must be chosen to help implement the program. It is essential that this entity can work as an intermediary between the funders and the community, and that it has fully embraced and internalized the concept, as well as the technicalities, of the program. An important argument for the intermediary being a nonprofit organization, as opposed to a government agency, is its ability to insulate itself from changes in the local political landscape.

4. *The program should have adequate financial and technical resources to do the job.* This should be self-evident, but, unfortunately, programs are too often undercapitalized at the outset or in making provisions for long-term management of the project or both. Along with this, it is important that programs earmark sufficient funding, so that tenant groups can obtain the technical assistance necessary to become full participants in a housing development-ownership-management process.

5. *The program must confront both the physical and social needs of a community.* Experience has shown that it is impossible and undesirable to think of housing strictly as a physical structure. Although the bricks and mortar must be tended to, the social environment of the buildings and the service needs of the residents must be acknowledged and addressed as well.

We have learned a great deal about how to do high-quality inner-city multifamily housing rehabilitation. In this sense, the cup is more than half full. On the other hand, the current mood in Congress does not favor large-scale government programs in housing or just about anything else. Even if top-down is emphatically coupled with bottom-up, the requirement of significant federal resources going into a program would likely be enough to kill any serious initiative. In this sense, the cup is much less than half full, and the future for low-income housing and other efforts aimed at improving life for poor inner-city residents looks bleak.

NOTES

1. Since these programs were launched, the Boston Housing Partnership merged with another nonprofit organization, Metropolitan Housing, Inc., in 1991 and changed its name to the Metropolitan Boston Housing Partnership. However, in this chapter, the original acronyms, BHP I and II, will be used. The name of the agency is referred to as either the BHP or the MBHP, depending on the particular time, before or after 1991.

2. The Section 221(d)(3) program authorized the federal government through the Federal National Mortgage Association (Fannie Mae) to purchase mortgages at below-market interest rates set at 3%. The lower interest rate resulted in reduced rents that were affordable by moderate-income households. Both for-profit and nonprofit developers were eligible sponsors. Preceding the 221(d)(3) program by two years was the Section 202 program that authorized direct federal below-market interest rate loans to nonprofit sponsors of elderly housing developments.

3. In the Section 236 program, loans were made to eligible sponsors by private lenders at market interest rates rather than directly by the federal government through Fannie Mae. The federal subsidy was in the form of a payment to lenders to reduce the interest rate down to an effective rate as low as 1%. As with the Section 221(d)(3) program, the reduced debt

service payments allowed owners to charge lower rents affordable by a moderate-income population.

4. These reports included the Report of the National Commission on Urban Problems (1968), the Report of the President's Committee on Urban Housing (1968), and the Report of the National Advisory Commission on Civil Disorders (1968).

5. Statement by Bryant Rollins, cited in Keyes, 1970, p. 2. Quoted from "Statement from Grove Hall Development Corporation et al.," December 4, 1967 (mimeo).

6. In addition to BHP and MHFA as the lead agencies, a host of additional city and state departments or agencies, as well as foundations, banks, and other financial entities were involved. (Metropolitan Boston Housing Partnership, n.d.[a]). The mayor's office was so fully supportive of the new initiative that it allocated $4.5 million of its Community Development Block Grant (CDBG) to the effort. This funding was absolutely essential for the program.

7. Although HUD was not a leader in the BHP I program, the initiative was still dependent on HUD subsidies in the form of the Section 8 program.

8. Letter from Thomas R. Bledsoe to Rachel G. Bratt, January 4, 1995.

9. Although BHP II officially only included about half of the package referred to as the Granite Properties (some 926 units), for the purposes of this chapter, it is reasonable to think of BHP II as being synonymous with the Granite Properties.

10. Letter from Thomas R. Bledsoe to Rachel G. Bratt, January 4, 1995.

11. Starting in the mid-1980s, subsidized housing developments that had been built under the Section 221(d)(3) and 236 programs reached the point (20 years after development) at which their private owners were, under their contractual agreements with HUD, entitled to pre-pay their mortgages and, in return, be relieved of any further commitment to rent the units to low- to moderate-income households. The possibility that many affordable units would be lost under this scenario is the essence of the "expiring use crisis." Federal legislation enacted in 1987 and 1990 attempted to safeguard these units, so that they could remain affordable. However, the cost of the programs has become the subject of considerable federal concern and scrutiny and future funding for the "preservation" program is precarious.

12. Author's interview, November 17, 1994, Boston, MA. Shortly after this interview, in early January, 1995, the top management team at MHFA, including Eleanor White, was replaced by new gubernatorial appointees. The new Executive Director, Steven Pierce, has gone on record strongly supporting Demo Dispo and has pledged its continuance along the lines of his predecessor, Marvin Siflinger.

13. The other three state housing finance agencies selected by HUD to implement Demo Dispo were Illinois, New York, and Washington, D.C.

14. It was also pointed out that "if the tenants don't perform, MHFA has the power to remove them from the center of the process and then we'll make the decisions for them." Author's interview, November 17, 1994, Boston, MA.

15. Letter from Thomas R. Bledsoe to Rachel G. Bratt, January 4, 1995.

16. A few weeks before this book went to press (July 1996), MFHA, in conjunction with the minority business community, agreed to change the 80% MBE goal to an 80% "community economic goal." This was prompted by recent U.S. Supreme Court decisions concerning affirmative action.

17. Interview with Robert Pyne, Director of Development, MHFA, March 7, 1995.

18. Author's interview, November 17, 1994, Boston, MA.

REFERENCES

Boston Globe. (1994, April 12). A brighter housing picture, p. 18.

Bratt, R. G., & Morris, E. J. (1989). HUD's property disposition policies and the Granite Properties. In R. G. Bratt, *Rebuilding a low-income housing policy* (pp. 146-165). Philadelphia: Temple University Press.

Bratt, R. G., & Plotkin, W. (1989). Institutionalizing community-based housing development: Case study, the Boston Housing Partnership. In R. G. Bratt (Ed.), *Rebuilding a low-income housing policy* (pp. 290-317). Philadelphia: Temple University Press.

Citizens Housing and Planning Association of Metropolitan Boston, Inc. (1971). *Preserving Boston's housing: An action program.* Boston: Author.

Conservation Company. (1991). *An evaluation of the Resident Resource Initiative.* New York: Author.

Justin Gray Associates. (1970). *Housing needs and priorities: Vol. I.* A report prepared for the City of Boston model city administration. Cambridge, MA: Author.

Keyes, L. C., Jr. (1970). *The Boston rehabilitation program: An independent analysis.* Cambridge, MA: Joint Center for Urban Studies of the Massachusetts Institute of Technology and Harvard University.

Massachusetts Housing Finance Agency. (1993). *Annual report.* Boston: Author.

Massachusetts Housing Finance Agency. (n.d., circa 1994). *Demonstration disposition information.* Boston: Author.

Metropolitan Area Planning Council. (1971). *Housing metropolitan Boston: The challenge and the response.* Boston: Author.

Metropolitan Boston Housing Partnership. (n.d.[a]). *Fact sheet on BHP I.* Boston: Author.

Metropolitan Boston Housing Partnership. (n.d.[b]). *Fact sheet on BHP II.* Boston: Author.

Metropolitan Boston Housing Partnership. (n.d.[c]). *Fact sheet on the Resident Resource Initiative.* Boston: Author.

Metropolitan Boston Housing Partnership. (1994, November). *Board report.* Boston: Author.

National Advisory Commission on Civil Disorders (Kerner Commission). (1968). *Report of the National Advisory Commission on Civil Disorders.* New York: Bantam.

National Commission on Urban Problems (Douglas Commission). (1968). *Building the American city.* Washington, DC: Government Printing Office.

Peterman, W. (1993). Public housing and resident management: A good idea gone wrong? *Shelterforce,* 15(6): 6-9.

President's Committee on Urban Housing (Kaiser Committee). (1968). *A decent home.* Washington, DC: Government Printing Office.

Siflinger, M. (1994, Spring). *MHFA Update: Vol. 13,* p. 4.

U.S. Department of Housing and Urban Development. (1980). *1979 statistical yearbook.* Washington, DC: Government Printing Office.

3 The Provision of Affordable Housing in Cleveland: Patterns of Organizational and Financial Support

NORMAN KRUMHOLZ

Abstract

This chapter traces the origins of Cleveland's neighborhood organizations from the settlement houses of the late 19th century through the protests of advocacy organizations of the 1960s to the nonprofit community development corporations (CDCs) of the present. It recounts the substantial accomplishments of the Cleveland Housing Network (CHN), an umbrella organization made up of 13 constituent neighborhood-based nonprofit housing organizations and now the most important producer of low- and moderate-income housing in Cleveland. It discusses intermediaries and funding sources crucial to CHN's success and considers issues important to the future of CHN and similar nonprofit housing providers in other American cities.

■

Over a period of little more than 100 years, the city of Cleveland grew from a single log cabin on the Cuyahoga River to one of America's mightiest industrial cities. The astounding growth of the population from 17,034 in 1850 to 796,841 in 1920 was a cause for self-congratulation among Cleveland's business leaders, but the city was being overwhelmed by the problems of congestion, poor sanitation, poverty, crime, and disease. As foreign immigrants flooded into Cleveland, they joined friends and relatives from the old country and created ethnic neighborhoods. Like poor people everywhere facing problems of unemployment, job discrimination, bad housing, and crime, they used these neighborhood organizations to improve aspects of their community and change the conditions that kept them poor and powerless (Van Tassel & Grabowski, 1986).

Background for Neighborhood Development in Cleveland

The first attempts at neighborhood reform sprang from ethnic church organizations and from the national trend known as "progressivism" or liberal reform. This movement became a nationwide political movement in 1900 with the succession of Theodore Roosevelt to the presidency, but it began in the 1890s with reform experiments on the local level. In Cleveland, this movement produced the landmark administration of progressive Mayor Tom L. Johnson (1901-1909) and the earliest settlement houses that were located in the city's most densely populated neighborhoods. The settlement houses tried to ease the burdens of the immigrant poor by providing valuable social services, such as education and homemaker training, but without challenging the distribution of money and power in the city. They could hardly do so, being supported by the charity of such millionaire industrialists as John D. Rockefeller.

The settlement house movement grew and prospered in Cleveland, and by 1990, its offspring, the Neighborhood Centers Association, had 25 agencies, a staff of over 600, and an annual budget of almost $14 million (Bond, 1990).

The liberal strategy of progressive reform represented by the settlement house movement depended on a sound and prosperous local economy. But although Cleveland's economy was rapidly expanding, economic conditions with respect to wages and working conditions were unstable. This resulted in a rash of labor troubles and a period of radical organizing that ran from the early years of the century through the Depression. During this period, independent labor organizers and socialists organized the unemployed in neighborhoods across the city.

During the Depression, Cleveland neighborhoods organized to support the growing labor movement. On December 28, 1936, 300 workers at the Cleveland Fisher Body plant, a division of General Motors (GM), initiated a sit-down strike at the Coit Road plant. The strike nearly paralyzed GM. During the strike, which lasted two months, organizations from the Collinwood neighborhood supported the sit-down strikers by passing in food to the strikers and holding supportive demonstrations. They persisted in their support, until GM recognized the United Auto Workers as the bargaining agent for its employees in February, 1937.

Following World War II, neighborhood organizing led by Cleveland and suburban neighborhood improvement associations took on a decidedly conservative turn. Most neighborhood improvement associations began as a means to lobby government for improved amenities and better coordinated local services. However, if the neighborhood was threatened with racial or class invasion as was common on Cleveland's east side, improvement associations often took on a decidedly protectionist and reactionary caste (Ross, 1955).

Urban renewal and the interstate highway program of the 1950s and 1960s foretold another phase of liberal neighborhood organizing. Interstate highways 77, 71, and 90 quartered westside neighborhoods like Tremont and cut huge swaths through other Cleveland neighborhoods. In the 1960s, Cleveland also mounted one of the nation's most ambitious urban renewal programs, certifying about 6,000 acres (one eighth of the entire city) for renewal (Keating, Krumholz, & Metzger, 1989). Urban renewal destroyed much housing and replaced relatively little; the city had wildly overestimated the market demand for its cleared land.

As the population of the city dropped sharply in the 1960s and 1970s because of deindustrialization, loss of jobs, and the mass exodus of the middle class to the suburbs, much abandoned housing was demolished by the city, often leaving behind trash-filled vacant lots. The public housing authority found it difficult to expand the stock of low-income housing, and many of the city's neighborhoods were redlined by commercial banks, savings and loans, and insurance companies, causing further disinvestment and decline. By 1990, the population of Cleveland was 505,000, slightly more than half of the 1950 population of 914,000. About 40% of all city families lived under the poverty line (Coulton, Chow, & Pandey, 1990).

Against this backdrop of general decline, community development corporations (CDCs) and advocacy organizations began to emerge, some as part of President Lyndon B. Johnson's social reform legislation. The Hough Area Development Corporation (HADC) was the first of these to be chartered in the 1960s. HADC was joined by the Model Cities Community Action Program in the late 1960s and the Famicos Foundation in the late 1960s. Among its other activities, Famicos pioneered an innovative model of lease-purchase homeownership for poor families that later became a mainstay program for Cleveland's emerging nonprofit housing corporations.

Other neighborhood groups, organized in Cleveland along the radical lines recommended by Saul Alinsky (1946) began to emerge with vigor in the early 1970s. Indeed, by the mid-1970s, Cleveland was undergoing a veritable revolt of the neighborhoods. The new groups were neighborhood-based advocacy organizations that focused on specific grievances and were confrontationist in style. They loudly insisted on an end to mortgage redlining, racial steering, and block-busting. They demonstrated for improved city services and the immediate demolition of abandoned and vandalized housing seen as a threat to their neighborhoods. They were also insistent upon getting a fair share for their neighborhood from the Community Development Block Grant (CDBG) (Bach, 1977). Some of these groups were administered and directed by the Commission on Catholic Community Action and partially funded by local foundations (Swanstrom, 1985).

These many neighborhood-based organizations, led with great vigor by neighborhood residents and with great talent by young staff organizers, seemed to hold considerable potential for more inclusive democratic politics in Cleveland and in the nation at large. Nationally, there were thousands of such advocacy organizations carrying on the legacy of resistance and insurgency and creating new forms of democratic participation and power sharing (Boyte, 1980; Fisher, 1984). There seemed to be excellent reasons to be hopeful about a blossoming of community-based organizing and development in Cleveland. However, in the late 1970s, the advocacy groups began to lose support, as they ran up against the inherently conservative nature of local business and politics. By the mid-1980s, virtually all of Cleveland's neighborhood-based advocacy organizations were defunct. However, almost all had spun off nonprofit housing development corporations, for the most part staffed by the talented, committed young men and women who had helped run the advocacy groups. These development-oriented CDCs took root, sought, and found necessary support to grow. The neighborhood movement in Cleveland had managed both to institutionalize some of its policies and to trim its sails to survive in the real world.

■ Cleveland Housing Network and the Lease-Purchase Program

The emerging CDCs needed tangible programmatic successes both to restore some of the tattered infrastructure in their neighborhoods and to ensure continued funding. They seized on the innovative lease-purchase model of the Famicos Foundation. During the 1960s, Famicos had developed a small program by buying vacant houses or accepting them as tax-deductible gifts, rehabilitating them to minimum city code standards, and leasing them at minimal cost to low-income families who, after 15 years of occupancy, could opt to take title to the property for the balance remaining on the mortgage. The program was simple, successful, and replicable. CDCs located and purchased homes from FHA or VA foreclosure lists or from private sources. Originally, the costs per unit were targeted at a maximum of $15,000 for acquisition and $10,000 for rehabilitation to city codes. Wherever possible, housing acquisitions were to fit in with the CDCs' plans for neighborhood revitalization.

In 1981, convinced that they could do better as a group and encouraged by local foundations, six neighborhood nonprofit housing organizations got together to form an intermediary umbrella organization called the Cleveland Housing Network (CHN). Originally, CHN was funded by the CDBG program, local foundations, and BP America. CHN was to facilitate and stimulate the housing development process by providing technical assistance and aiding in the financing and managing of lease-purchase units. In addition, CHN pioneers believed an umbrella organization could locate more funders for the CDCs and more sponsors for low-income housing tax credit (LIHTC) syndications[1] than individual CDCs. Over the next 10 years, CHN, with the support of key institutions in the public and private sectors, added seven affiliate CDCs for a 1993 total of 13. CHN's production of lease-purchase units has steadily increased from 20 units a year in 1981 to 210 units in 1994. By 1993, the CHN lease-purchase program was the oldest and largest program of its type in the country.

In terms of process, a qualified lease-purchase family is selected for each housing unit. Most families have very low incomes. Since 1986 when the LIHTC provided an additional funding source, the program has been directed at the very poor with 90% of participating households having incomes under 30% of the median. CHN's "average" family is headed by a single woman with three children living on less than $8,000 a year. Each family goes on a waiting list then passes through a screening process, including references, credit check, and a home visit, receives training in home maintenance, and takes responsibility for basic maintenance on the house upon occupying it (Clint & Wertheim, 1979). Lease-purchasers pay "rent" in the $150 to $200 a month range, depending on the size of the unit, so that families whose only income is welfare can afford some units.

Family social services are also made available to the tenant through a Family Development Program that focuses on addressing the problems of long-term poverty among CHN's lease-purchase families. The program is designed to help these low-income families connect with services, find opportunities for growth, rise out of poverty, and build a better future. It has been quite effective (Balfour & Leahy, 1993).

At the end of 15 years, when the low-income housing tax credits are exhausted, if the family has remained a tenant in good standing and participated in homeownership training programs, then the family may take title to the property for the outstanding indebtedness. Once a family takes title, there are no restrictions to ensure that the unit remains affordable. Households that progress to homeownership inherit a first mortgage of about $6,000. Of all the programs in the CHN arsenal, lease-purchase has been the most popular with tenants as well as public and private participants. The program always has a waiting list that is two or three times longer than the number of units available (in 1994, the waiting list was more than 200 families), an annual turnover rate of less than 3%, and a seven-year cumulative turnover rate of less than 20%.

■ The Homeward Program

In addition to its lease-purchase program, CHN also provides a more immediate homeownership program. CHN's Homeward program began in 1989 to stimulate owner-occupied single family housing. By 1994, CHN had acquired and rehabilitated over 250 houses. These have been sold conventionally to low- and moderate-income families at an average price of $50,000. CHN has negotiated discounted first mortgage financing from local lenders at interest rates that are significantly below the market.

The Homeward program also offers a deferred $5,000 second mortgage to each buyer that is financed and held by the City of Cleveland's CDBG funds. These funds are allowable only for families who earn less than 80% of the Cleveland area median income. Construction-period working capital is a combination of subsidized funds from local foundations and banks.

Houses are sold to financially qualified families on a first come-first served basis. Potential buyers are prescreened by CHN's affiliate member groups, and to date 75% of all buyers have been minority families. CHN offers its member groups ongoing training in prequalifying buyers and underwriting and credit standards.

For the lease-purchase and Homeward programs, CHN owns the units in fee and finances their rehabilitation, while the affiliate CDCs develop and manage the properties. The CDCs are responsible for the acquisition, site preparation, and contracting, while CHN provides financing and technical assistance. The CDCs also screen and select the households who occupy the units and collect rents.

Property management has been a continuing although, so far, low-level problem. The rehabilitated properties, after all, are about 90 years old and were originally built as modest, wood-frame, working-class housing. To forestall a larger maintenance problem in the future, CHN intends to centralize management and assume property maintenance and repair. With CHN handling maintenance, the affiliate CDCs should be able to increase their development and their tenant screening and counseling roles.

CHN is also attempting to realize economies of scale and raise production levels among CDCs by awarding development fees based on levels of production. To receive any development fees, a CDC must produce at least 15 units a year. The more units produced by a CDC, the more incentive development fees it receives. It is hoped that such incentives will lead to higher levels of housing production that in turn will lead to a better chance to revitalize disinvested communities.

CHN offers other programs and has also provided support to member CDCs for the acquisition, rehabilitation and management of small, multifamily apartments. Finally, CHN's CDCs have used the Federal Home Weatherization Assistance Program (HWAP) and the Housewarming Program funded by the local utility company to weatherize over 25,000 units of housing.

By 1993, CHN had compiled an enviable record. The organization had assisted in the acquisition and rehabilitation of about 1,000 housing units in Cleveland. Of the total, about 800 were lease-purchase units, 72 were rental units in multifamily projects, and about 250 were Homeward units. An additional 180 units were scheduled for completion in June, 1994. Total production of all housing of all CHN affiliates is now about 320 units a year, up from 200 units a year in 1991. In the last three years, total production has increased by 60%.

CHN had a 1993 professional staff of 40 and an annual operating budget of $2.2 million; $1.6 million of the total amount goes to fund CHN's internal operations and $600,000 is passed-through to affiliate CDCs in the form of operating support, management fees, and development fees. The sources for CHN's funding are broken down as follows: 30% from corporations and foundations, 15% from weatherization contracts from the public and private sectors, 15% in other government contracts, and 40% in the form of management and development fees.

Control and Capital

One of the major criticisms of previous urban redevelopment efforts was that the people who lived in the neighborhoods affected by the programs had no voice in their planning and implementation. So far, this has not been the case at CHN. Even though CHN targets low- and moderate-income households, the organization does not implement plans or targeting approaches in a top-down manner. On the contrary, policy percolates in general from the bottom up. Two important cautionary notes should be added, however. First, in some neighborhoods, property owners and investors take an active role on the CDC board and sometimes try to skew priorities away from affordable housing. Second, although it happens infrequently, when the mayor or the ward councilperson forcefully expresses a priority, the CDC is inclined to go along.

A majority of CHN's board of directors is drawn from representatives of the affiliate CDCs. Their boards, in turn, are made up of residents and businesspeople from their neighborhoods. The affiliate CDCs decide themselves on the mix of low-, moderate-, and middle-income housing based on debate among board members as well as funding and development opportunities. CHN, however, insists on a minimum number of lease-purchase units each year as a condition of membership. CHN's planning process seems fully responsive to neighborhood needs and desires, although the city's overall housing policies are considered as well.

Financial capital for CHN's lease-purchase program comes from a variety of sources, both public and private. The city of Cleveland provides both interim and permanent debt through its rental rehabilitation and weatherization programs and through CDBG funds. The state of Ohio allocates low-income housing tax credits and energy conservation grants through the HWAP program for the projects. Local commercial banks and S&Ls provide private debt with linked deposits[2] made by church or philanthropic agencies seeking social returns by subsidizing the loan. The Enterprise Foundation supports CHN by providing working capital. Equity investments for the projects come from Cleveland's corporations through the Cleveland Housing Partnership Equity Fund. Family Development Programs come from the George Gund Foundation and the Ohio

Center for Family Development. Clearly, CHN's sources of financial support are highly diversified.

Several corporations with local headquarters invest once a year in the Cleveland Neighborhood Equity Fund, seeking low-income housing tax credits. The fund is a limited partnership managed by the Enterprise Foundation, the Baltimore-based nonprofit that pioneered the public-private partnership concept. The CHN lease-purchase program is the single largest project invested in each year, although other low-income housing rehabilitation projects are also targeted. In 1991, $5.5 million was invested in the CHN lease-purchase program, including the rehabilitation of single and two-family homes in selected neighborhoods. Because the properties are rented as low-income housing, they earn tax credits that are then syndicated back to the original corporate investors—which, therefore, can count on annual returns of 15% or more.

The complexity of the process can be appreciated by looking at the "patchwork" financing of a typical CHN housing package. Sources of investment might include but are not limited to the CDBG program, an allocation to the city from the federal government; other city and state funding; the Low-Income Housing Tax Credit; private and philanthropic loans or grants; mortgage insurance from the Federal Housing Administration; bank financing for construction funds and permanent mortgages; city tax abatement; and so on. Stitching together this patchwork is no small accomplishment.

Important Intermediaries

City of Cleveland

The city of Cleveland is CHN's most important source of support, although its neighborhood strategy is sensitive to political considerations that attempt to satisfy all housing constituencies. The current administration has substantially increased its support for neighborhood-based housing.

Cleveland's Housing Trust Fund, established in 1991, which the city controls and funds out of its CDBG allocation, is an example of this broad approach to housing. The city has divided its $4.4 million trust fund into two equal streams; one funds new construction and substantial rehabilitation projects, while the other funds low-income rental and lease-purchase projects. The maximum award available under either stream is

$500,000 that may fund no more than 25% of new construction projects and no more than 50% of low-income housing projects.

In addition to trust fund monies, the city provides substantial funds from the HOME program in the form of below-interest construction loans for CHN's lease-purchase program. The city also provides periodic infusions of CDBG funding to the Cleveland Action to Support Housing (CASH) program, a 15-year-old rehabilitation program that leverages private financing and helps low- and moderate-income homeowners to repair their units.

There are two other important CDC-based funding streams from the city in support of neighborhood housing. The first allocates $1.5 million a year for core operating support for CDCs through a competitive process. Local CDCs submit proposals to the Community Development Department that awards grants between $45,000 and $120,000 to organizations that meet certain production benchmarks. The second funding stream may be unique to Cleveland among all U.S. cities receiving CDBG funding: Each of Cleveland's 21 councilpersons is allocated $300,000 annually that may be distributed almost entirely at his or her own discretion. In fact, many councilpersons award much or most of their allocation to neighborhood CDCs for housing improvements.

Neighborhood Progress Incorporated (NPI)

While CHN was a product of grass-roots evolution, NPI was a creature of the foundation and corporate communities. Three pivotal actors in the formation of NPI were the Cleveland Foundation (1994 assets of $750 million), the George Gund Foundation (1994 assets of $430 million) and Cleveland Tomorrow, a consortium of 50 of the largest corporations in the Greater Cleveland area. These actors believed that CHN's program could be assisted by a technical and financial assistance intermediary like NPI that could channel corporate and foundation support in a way designed to maximize neighborhood revitalization. NPI, it was felt, could force increased CDC production more easily than CHN.

Accordingly, NPI was established in 1988. Its initial task was to prepare a strategy for neighborhood investment in Cleveland and encourage CHN and its members to "build projects of scale" and raise their production levels. NPI's still evolving role includes assigning priorities to the funding of neighborhood programs and stimulating corporate involvement in those programs. In the 1993-1994 budget cycle, NPI had a grant budget of $2.8 million. Its loan fund as of June 1993 was capitalized at $2.5 million.

NPI provides funding to CHN and CDCs and collaborates with them on development projects. For example, NPI's Cleveland Neighborhood Partnership Program competitively awards core grants of around $30,000 over a two-year cycle to CDCs. CHN has received over $500,000 in project grant support between 1991 and 1993. NPI subsidiaries also provide debt financing for development projects and enter into partnerships with neighborhood groups on large and complicated projects. NPI subsidiaries receive about $4 million a year from Cleveland Tomorrow's Community Development Program, and NPI also provides technical assistance to CDCs and plays a crucial role in funding the operating budgets of two other important local intermediaries, the Local Initiatives Support Corporation (LISC) and the Enterprise Foundation.

Local Initiative Support Corporation (LISC)

LISC established its local operations in Cleveland in 1982. Its primary interests are moderate-income housing and commercial development for which it provides technical and financial assistance. The agency tends to defer to NPI in overall policy direction, which is not surprising, because NPI provides its operating support of about $100,000 a year. NPI also provides LISC with local loan funds. From 1982 to 1992, LISC made about $4.5 million in grants and development loans, with LISC's national office and local contributors, such as the Cleveland Foundation, BP America, and the Gund Foundation, providing the local share. In 1989, LISC loaned about $1 million to local CDCs.

Enterprise Foundation

The local Enterprise Foundation office in Cleveland is particularly useful to CHN because of Enterprise's expertise in raising equity through syndications. Begun in 1988, Enterprise's local office has a two-person staff and an annual operating budget of $120,000, 70% of which is from syndication fees and the balance from local foundations and corporations. Enterprise manages the Cleveland Housing Partnership Equity Fund (CHPEF) in behalf of the investors from Cleveland Tomorrow. It also provides technical assistance to CDCs on property management, accounting, organizational development, and housing development.

Enterprise's most important role has to do with the management of the CHPEF. Utilizing the Low-Income Housing Tax Credit, Enterprise raises equity for CHN projects. From 1986 to 1992, $17.4 million in equity had been raised from 15 projects involving about 1,000 units of low-income housing.

Enterprise concentrates its efforts on low-income housing. Although the organization recognizes that low-income housing may only be one part of a comprehensive revitalization strategy, it realizes that this part is most easily overlooked or diluted. Accordingly, Enterprise tries to ensure that the low-income housing component remains an important part of the city's redevelopment strategy.

State of Ohio

The state directly allocates substantial sums to Cleveland's CDCs through the Department of Development and the Ohio Housing Finance Agency. The Department of Development offers technical and financial assistance, such as development grants, loans, and capital funding to nonprofit organizations in the state. Ordinarily, the state's subsidies will finance somewhere from 10% to 15% of the cost of a CDC's developments. Sources of revenue for the state subsidies include general revenues and a statewide housing trust fund supported by state-issued bonds. The Development Department also provides grants to match contributions by the East Ohio Gas Company to assist CHN in administering the weatherization program.

In addition to the Department of Development, the Ohio Housing Finance Agency (OHFA) issues mortgage revenue bonds (MRB) that can be used to generate below-market mortgages for single-family housing. In 1990, OHFA contributed $4 million in mortgage financing for CHN's Homeward program. Recently, however, MRB financing for CHN's projects has declined by about half, because CHN can acquire more favorable financing on the private market.

The Ohio Community Development Finance Fund (CDFF) is a quasi-public institution with an independent board providing support for CHN and CDCs. CDFF has been involved since 1990 in a variety of Cleveland housing developments with a total value of $11.5 million. Ordinarily, CDFF combines its funds, derived from state general funds, with contributions from NPI or a local church. These monies then leverage below-market-rate mortgages for CHN lease-purchase projects. CHN captures a good share of CDFF's funds, because CHN is perceived as a well-established and competent housing provider.

Local Foundations

From the earliest days of the neighborhood movement, Cleveland's foundations have played a critical role. The Cleveland Foundation and the George Gund Foundation are philanthropic pacesetters. Their support for neighborhood advocacy organizations in the 1970s led directly to backing for the CDCs and CHN. Among CDCs on their long list is virtually every CDC and neighborhood group in the city of Cleveland. The two Foundations have also extended their largesse to a host of other organizations supporting housing and community development, including LISC, Enterprise, NPI, and the Center for Neighborhood Development at Cleveland State University.

The foundations have assisted CHN and its CDC affiliates in the undertaking of a housing job that the unassisted private market has shown no interest in tackling. It seems clear that CHN has been successful and has in fact been instrumental in restoring the confidence of many individuals residing in Cleveland's neighborhoods and doing business there and in stimulating the interest of state and additional funding parties outside the area. CHN's production record is impressive and growing stronger; its support from foundations, government, and business seems very solid.

Five Reasons for CHN's Success

1. The CDCs. The neighborhood advocacy efforts in Cleveland of the late 1960s and 1970s brought a remarkable group of young men and women to the fore. Committed to helping low-income and working-class people and neighborhoods, these people remained in Cleveland, continued working, and survived the shift from advocacy to development. Fifteen years later, they make up a competent and supportive network of CDC directors and staff, city hall officials and politicians, and bank-lending officers. As the nonprofits moved away from advocacy and into development in the early 1980s, their ability to successfully rehabilitate housing against great odds was quickly demonstrated. It proved the shortest path to community credibility. Without such standing, a CDC can accomplish little of lasting value.

Once the fledgling CDCs completed their first housing projects, their survival depended on maintaining their neighborhood and institutional support. This required the ability not only to produce housing but to manage it efficiently and attract resources in a sustained flow. Management, in turn, involved issues of tenant screening and selection, rent collection and eviction, as well as dealing with delinquency, drugs, graffiti, and the mountain of other problems that are present in neighborhoods of concentrated poverty and deprivation. CDC leadership and capacity grew with each challenge.

Cleveland neighborhoods now have in their CDCs a new institutional force that was not present 15 years ago. That force has an unusual combination of entrepreneurial, management, and political skills and the perseverance to carry forward with their programs against great odds. It may be that the CDCs will provide the leadership, not only to rebuild the housing of the community but to rebuild its shattered social and economic structure as well.

2. The Philanthropies and Intermediaries. As has been noted, the support of local philanthropies, especially the Cleveland Foundation and the George Gund Foundation, has been essential to the development of the neighborhood nonprofit housing corporations in Cleveland as well as the earlier neighborhood advocacy groups. Without their early support for staffing and planning, it is doubtful the neighborhood movement could have developed to its present extent.

Another important reason for CHN's success is the rise of local intermediaries to raise capital, provide technical assistance, and generally improve CDC capacity. The most important of these intermediaries are the Local Initiative Support Corporation (LISC) and the Enterprise Foundation. Both of these organizations locally replicate functions performed by their national counterparts. LISC was established in 1979 by the Ford Foundation to leverage contributions by corporations and local philanthropies in support of CDC activities. Enterprise, established by developer James Rouse in 1984, raises private capital to build low-income housing and strengthen nonprofit housing corporations.

These two intermediaries have performed three essential functions in Cleveland: (a) They have helped mobilize equity capital and assisted CDCs in financial packaging; (b) they have provided technical assistance in project development, especially with respect to tax syndication; and (c) they have improved the technical competence of CDCs and made it easier for banks and other funders to accept the legitimacy of the CDC and deal with them. To an extent, the work of these intermediaries has offset some of the problems caused by undercapitalization and the "patchwork" financing of neighborhood projects.

3. Community Reinvestment Act (CRA).[3] Another factor in the success of CHN is the CRA and the willingness of the city to use the act to pressure lenders to extend credit in formerly redlined and disinvested neighborhoods. In the 1960s and 1970s, many commercial banks, savings and loans, and insurance companies denied credit and insurance in some urban neighborhoods, a process known as redlining. Advocacy groups and their allies complained that these redlined neighborhoods were unfairly locked into a self-fulfilling prophecy of neighborhood decline. In response, Congress enacted the Home Mortgage Disclosure Act of 1975 and the Community Reinvestment Act of 1977 that allowed community groups and cities to challenge lenders on whether they are providing adequate credit in certain neighborhoods. Challenges must be reviewed by the lenders' regulatory agency. The process can take much time, making the lenders anxious to avoid a challenge. CRA legislation was strengthened by Congress in 1988.

The current Cleveland administration has been aggressively using the CRA to negotiate commitments from lenders to provide credit in all city neighborhoods. Agreements totaling hundreds of millions of dollars have been reached, and disinvested neighborhoods and their CDCs are being sought out by lenders anxious to make deals for the first time.

4. Low-Income Housing Tax Credit (LIHTC). Some form of the LIHTC is essential for the continued success of CHN as a developer of low-income housing. LIHTC offers corporations an opportunity to invest in low-income housing, while at the same time receiving a reasonable rate of return on their investment. The direct tax write-offs on corporate tax liability made possible by LIHTC now pay off about 55% of the equity capital—that is, invested rather than borrowed money—in most CHN projects. In the United States at large, LIHTC has become the primary subsidy tool for low-income housing production. It was permanently renewed by Congress in 1992.

5. Land Banking. To a certain extent, laws put into place in the 1970s as a reaction to major problems of tax delinquency and property abandonment have been useful to the CDCs in the 1990s. One example is Cleveland's Land Bank that won a Harvard Kennedy School award in 1994 for urban innovation. The land bank was begun in 1975 as a means of recapturing tax-delinquent properties, clearing the titles, and recycling them back to productive use (Krumholz & Forester, 1990). Many CDCs have made good use of land-banked properties that are cheaply available as sites for new housing. This has been especially true in the Hough neighborhood, where rental apartments and expensive single homes have been developed using land-bank properties. However, there are still serious problems in Hough, including widespread poverty, large vacant land areas, and decayed factory buildings beyond the scope, so far, of CDC activities.

Remaining Problems

However effective and successful CHN has been, important issues and problems remain. The most important of these is the need for continued strong financial and political support. The November 1994 elections, which produced budget-cutting majorities in the House and Senate, threaten that support. One can only hope that Congress will continue to sustain the CDBG and HOME programs at current funding levels. CRA must also remain, because its mandates make it possible for cities like Cleveland to pressure commercial banks and savings and loans to lend in heretofore disinvested neighborhoods. Cleveland's present administration has negotiated several important agreements with local banks along this line with CRA as the lever.

Nevertheless, as the 1994 elections remind us, long-term financial and political support for CHN and the CDCs is by no means certain. The permanent renewal by Congress in 1992 of the low-income housing tax credit (LIHTC) was an important commitment to a major subsidy, but it can be rescinded. State government, faced by fiscal austerity, may cut subsidies to local CDCs. Foundation support, mostly from the Cleveland and Gund foundations, has been particularly crucial in Cleveland for the creation of CHN and its intermediaries; yet, the long-term commitment by foundations to such enterprises may still prove to be fickle. Corporate philanthropy is not a predictable constant but will rise and fall with the fortunes of the companies involved and their changing strategies for giving.

Critics have also pointed to other problems. CHN, like many of its affiliates, frequently carries out development projects on an ad hoc basis, without having an overall neighborhood revitalization strategy. Some projects are selected because of opportunity, price, and availability rather than their "fit" into an overall plan. Most CDCs feel they can succeed by addressing problems only within their own neighborhoods, but many neighborhood problems did not originate there and cannot be resolved there.

It is important that CHN activities be linked to a more comprehensive city neighborhood improvement strategy, but the neighborhood groups cannot be expected to do it. Neighborhood groups are partisans for their own turf. That makes them ideal organizations to pressure the city to improve services, major institutions and businesses to support their neighborhood, and property owners to maintain standards. A long-term, comprehensive neighborhood revitalization strategy is essential, but it must be developed by the city.

Another serious problem is the possible relegation of low-income housing production to a lower priority than it has enjoyed up until now. There is emerging evidence that the commitment of some of the CHN neighborhood-based affiliates to the production of low-income housing has eroded. Both the city and NPI have indicated growing interest in market-rate housing for middle-class buyers. The mayor has announced a goal of building 1,000 new homes each year and NPI supports more emphasis on market-rate housing. Observers believe that "many neighborhood groups are losing their commitment to low-income housing in favor of market-rate housing" (Chatman, 1994). On Cleveland's west side, a CDC that has been one of the most effective producers of lease-purchase housing has had its area of operations sharply limited by its councilpersons and has lost the support of NPI. This CDC's future is very much in doubt. An eastside CDC, also a major lease-purchase producer, has now focused its attention on building a 215-unit "suburban" subdivision with home prices starting at $85,000. Other CDCs report their rising concern about "deteriorating absentee-owned rental units" and the need to emphasize homeownership at market rates. In such traditional black eastside neighborhoods as Hough and Glenville, each with 1993 poverty rates of over 50%, CDCs are beginning to emphasize the construction of new, heavily subsidized single-family detached homes for bankable families. In 1993, half of Cleveland's $4.4 million Housing Trust Fund went to subsidize some of these market-rate new homes ranging in price from $85,000 to $275,000, while the mayor is pushing hard for 1500 new units of heavily subsidized housing in downtown (Chatham & Lubinger, 1994). Meanwhile, the 1993 median price for all homes sold in Cleveland was $47,600.

There may be several reasons for this shift in priorities from low-income housing to higher-income properties. First, in a very poor city where 1993 poverty rates were over 40%, Cleveland's official housing policy tries to serve the poor but also puts heavy emphasis on bringing the middle class back to the city. Cleveland's foundations, corporate community, and NPI all support this upscaling strategy. Because CHN support depends on these actors, it cannot but go along. Second, the CDCs may be captured by the "ideology of property ownership" (Fainstein & Hirst, 1996). Components of this ideology include the following beliefs: Owners have a stake in the neighborhood and in the high maintenance of their property, while renters are transient and care little about either the neighborhood or property. In addition, concentrations of renters (or low-income leasees) raise levels of poverty, bring attendant social problems, and inevitably spell neighborhood decline. Therefore, if the goal is neighborhood improvement, then one should provide benefits for property owners, while actively discouraging renters. Also, from the CDC's perspective, it is easier to build new housing for sale or rehabilitate and sell a Homeward unit than to manage a lease-purchase property for 15 years, while also dealing with any tenant problems.

Another possible reason for the shift in priorities, as previously noted, may reflect the fact that persons of higher income and owners of property are more likely to participate on the boards of CDCs than low-income renters (Berry, Portney, & Thompson, 1993; Thomas, 1986). Hence, the board members of CDCs may be more interested in accumulation and the protection of property values than with providing decent housing for people who need it most.

Finally, CDCs may be struggling with racial issues compounded by the dilemma of targeting local residents versus open access. By definition, CDCs want to target improvements to local residents, but various legal regulations insist on open access. Participants in CHN's programs are simply taken from qualified waiting lists. Most participants in the lease-purchase and Homeward programs have been minorities. Because all CDCs are territorially based and many neighborhoods in Cleveland are racially isolated, attracting "outsiders," especially those who are members of a different racial group, is often controversial. As an example, when the Broadway Area Housing Coalition, a CHN member, attracted African American tenants to its subsidized housing, some tenants met with racial harassment and complained to Cleveland's Community Relations Board. Most of Cleveland's politicians have shied away from promoting racial diversity, and Cleveland is one of the most racially segregated cities in the country (Massey & Denton, 1989). To overtly promote racial diversity, CDCs will need the strongest kind of support from all parts of the local and larger community, including the federal government.

Other problems related to CHN and its housing affiliates have to do with the gap between the stated goals of the "neighborhood movement" and the real accomplishments of the Cleveland groups. The CDCs have neither attained the goals of neighborhood empowerment of the poor or neighborhood self-reliance, nor have they been able on their own to transform and uplift the troubled poor and working-class neighborhoods of the city. To a degree, these more radical, reformist goals have been held hostage by the practical need to work within the existing private housing market and seek support where it could be found. As a result, few of CHN's affiliates have been involved in economic or job development projects or comprehensive neighborhood planning. They have been largely apolitical, depending for their support on the generosity of public, private, and philanthropic funders who make their housing developments possible and whom they cannot afford to alienate.

What has evolved in Cleveland is a small cadre of community housing developers, supported by a web of personal and professional relationships among public agencies, private funders, corporate philanthropies, and local government. The result has been a remarkably effective network of CDCs under CHN. They generally remain socially conscious in their goals, and they provide most of their benefits to low-income people. With any stagnation or loss of funders' interest, the relationship may begin to unravel. In that event, long-term prospects for continued success could become clouded. Such problems could end the bid of agencies like CHN in cities around the country to continue as major players in the revitalization of America's urban areas.

NOTES

1. Syndication process: Each project done with Low-Income Housing Tax Credits is allotted a certain amount of tax credits. Ownership is structured as a limited partnership, so that corporate investors can "buy" tax credits that are a direct write-off on corporate tax liability. Such credits pay corporate investors about a 15% rate of return and pay for about 55% of most CHN projects.

2. Linked deposits: deposits by church congregations or philanthropies in lending institutions. The depositors agree to accept a low rate of return on their deposits, while the lender agrees to make loans to neighborhood-based housing projects that may have relatively high risks.

3. CRA: In 1977, Congress passed the Community Reinvestment Act into law. CRA allows a city or neighborhood organization to challenge the proposed action (merging, closing a branch, and so on) of any commercial bank or savings and loan. If the challenger files a complaint, it must be investigated by the lenders' regulation. This stops or delays the proposed action. Because time *is* money, lenders are anxious to avoid challenges and may be willing to commit to substantial neighborhood lending if the complaint is forestalled.

REFERENCES

Alinsky, S. D. (1946). *Reveille for radicals.* Chicago: University of Chicago Press.

Bach, V. (1977). The new federalism in community development. *Social Policy, 7*(4): 32-38.

Balfour, D. L., & Leahy, P. S. (1993). *Process evaluation for the Ohio Center for Family Development.* Unpublished paper, University of Akron, Department of Public Administration and Urban Studies.

Berry, J. M., Portney, K. E., & Thompson, K. (1993). *The rebirth of urban democracy.* Washington, DC: Urban Institute Press.

Bond, R. L. (1990). *Focus on neighborhoods.* Cleveland: Greater Cleveland Neighborhood Centers Association.

Boyte, H. C. (1980). *The backyard revolution: Understanding the new citizen movement.* Philadelphia: Temple University Press.

Chatham, A., & Lubinger, W. (1994, June 29). City breaks ground for first apartments in downtown in twenty years. *Plain Dealer,* p. 1.

Chatman, A. D. (1994, May 7). He advocates housing for the poor. *Plain Dealer,* Real Estate Section, p. 31.

Clint, E., & Wertheim, K. (1979, November). How small-scale rehab works in a northern inner city. *Planning, 45*(11): 17.

Coulton, C. J., Chow, J., & Pandey, S. (1990). *An analysis of poverty and related conditions in Cleveland area neighborhoods.* Cleveland: Center for Urban Poverty, School of Applied Social Services, Case Western Reserve University.

Fainstein, S. B., & Hirst, C. (1996). Neighborhood organizations and community planning: The case of the Minneapolis neighborhood revitalization program. In W. D. Keating, N. Krumholz, & P. Star (Eds.), *Revitalizing urban neighborhoods.* Lawrence: University Press of Kansas.

Fisher, R. (1984). *Let the people decide: Neighborhood organizing in America.* Boston: Twayne.

Goetz, E., & Sidney, M. (1994). Revenge of the property owners: Community development and the politics of property. *Journal of Urban Affairs, 16*(4): 319-334.

Keating, D., Krumholz, N., & Metzger, J. (1989). Cleveland: Post-populist public-private partnerships. In G. D. Squires (Ed.), *Unequal partnerships.* New Brunswick: Rutgers University Press.

Krumholz, N., & Forester, J. (1990). *Making equity planning work.* Philadelphia: Temple University Press.

Massey, D., & Denton, N. (1989). Hypersegregation in U.S. metropolitan areas. *Demography, 26*(3): 373-391.

Ross, M. G. (1955). *Community organizing and planning.* New York: Harper & Bros.

Swanstrom, T. (1985). *The crisis of growth politics: Cleveland, Kucinich, and the challenge of urban populism.* Philadelphia: Temple University Press.

Thomas, J. C. (1986). *Between citizen and city: Neighborhood organizations and urban politics in Cincinnati.* Lawrence: University of Kansas Press.

Van Tassel, D. D., & Grabowski, J. J. (1986). *Cleveland: A tradition of reform.* Cleveland: Kent State University Press.

4

Twenty-Five Years of Community Building in the South Bronx: Phipps Houses in West Farms

LYNDA SIMMONS

Abstract

West Farms, the Bronx, New York City, is a real, although beleaguered, community in 1996. Twenty-five years ago, it was virtually an empty cocoon, its residents forced out by construction of the Cross Bronx Expressway, which destroyed its buildings and social fabric. Subsequent years brought physical and social changes often associated with war: destruction, chaos, and population migration. In West Farms, however, the changes resulted from governmental actions and policies. Phipps Houses (PH), the nation's oldest not-for-profit housing and community development corporation, entered West Farms to help build homes and redevelop a community; over these years it found a workable way of dealing with inner-city problems. In the process, PH was itself transformed, as its educational and social service arm, Phipps Community Development Corporation, and the property management subsidiary, Phipps Houses Services, Inc., came to play an equal role with the parent development organization. In the early 1990s, the cumulative work began to bear fruit. This chapter is the brief story of those 25 years.

■

AUTHOR'S NOTE: This chapter is a condensation of a more detailed account, which may be obtained by writing to Phipps Houses, 43 West 23rd Street, New York, NY 10010 (Attn: President's Secretary).

History and Background

West Farms is a 27-square-block area in the Bronx, New York. Its earliest settlers were farmers; by the 1890s it was a riverport village. The completion of the subway at West Farms Square in 1906 spurred residential development for the workers of Manhattan. In the 1920s, there was no ground left, except streets and a small cemetery. Six-story apartment buildings became home to lower-paid workers of the downtown garment industry. The neighborhood became a launching area for upwardly mobile working-class immigrant families. West Farms was a vibrant and thriving neighborhood into the 1950s.

It was first destroyed in the 1950s by Robert Moses's Cross Bronx Expressway. The Cross Bronx was only one of a series of governmental actions that, until the 1990s, broke down the community and its physical setting and thwarted local redevelopment efforts. The expressway demolished a blockwide swath of buildings and forced relocation of hundreds of families from the large rent-controlled apartments. The six-lane river of cars, noise, and pollution fractured the old East Tremont area.[1]

The population declined and its composition changed drastically. This exchange of populations in West Farms, between about 1955 and the late 1970s, can only be compared with mass migrations induced by war, disease, or economic depression. Older Whites moved out and younger Blacks and Hispanics moved in. By 1970, 14,000 (57%) of the Whites in West Farms in 1940 had moved out. There were nearly twice as many Blacks as Hispanics (around 6,200 to around 3,300), the Black portion having nearly doubled between 1960 and 1970. Hispanics had been too few to be reported separately in the census before 1970, but by 1980 they were about 60% of the population, reversing political and social forces in the neighborhood. By 1990, they accounted for 44% of the Bronx population and 72% of West Farms' (see Table 4.1).

Housing officials of New York City were relocating poor people displaced by urban renewal into the large, inexpensive (and dilapidated) apartments of West Farms. Political district boundaries ran through West Farms, putting it at the tail end of several districts; few politicians cared much about it. Rent control, never lifted after World War II, had diminished funds for buildings' upkeep, and after the destruction wrought by the Cross Bronx, many landlords reduced or ceased maintenance. In 1963, West Farms was declared an area of "deteriorating factors and conditions"[2] and an urban renewal plan was created.

TABLE 4.1 West Farms Population, 1940-1990

Year	*White*	*Black*	*Hispanic*	*Other*	*Total*
1940	24,581	33	N/A	7	24,621
1950	21,973	231	N/A	16	22,220
1960	18,101	953	N/A	57	19,111
1970	10,614	6,191	3,300	685	20,790
1980	166	4,343	7,060	120	11,689
1990	61	3,371	9,198	127	12,757

NOTE: Census data from inner-city areas present problems in that they are often internally inconsistent. Totals and subtotals in different sections of the census tract reports may not add up, some data may be missing for some tracts, and data collected in different years may vary in category definition, preventing direct comparisons from year to year. Census Bureau demographers, for instance, discount the 1970 figures on Hispanics because their own categories were ill defined. In addition to inevitable human error, census takers may make up (construct) data as best they can when not admitted to homes or they fear to visit certain buildings or blocks. Some people surveyed do not tell all the truth for fear of losing welfare or housing subsidies. In the case of data presented here from Bronx Census Tracts 359, 361, and 363, I have analyzed the data extensively to interpolate sensibly (on the basis of personal knowledge of the time) to supply Hispanic figures for 1970. I am also indebted to West Farms Planning Task Force (1994).

In 1969, the East Tremont Neighborhood Association (ETNA) asked Phipps Houses (PH) to come in and sponsor the major development in the Bronx Park South Urban Renewal Area. ETNA wanted PH, which was then virtually the only not-for-profit housing development organization in the city and had a long record (back to its founding in 1905) of successfully developing and managing housing for low-income households (see Figure 4.1). It also had funds to finance the substantial equity required of all developers, including not-for-profits, until the 1969 Tax Reform Act created tax shelters as a means of getting equity into low-income developments. Its oldest existing community in Sunnyside, Queens, is still visited by architects and planners from around the world.

During its involvement in West Farms, PH went from 6 staff in the central office and 40 or 50 in the field in 1969 to around 60 central staff and 540 field staff in 1994. Its 1995 core budget was over $11 million. Budgets for its own properties add another $31 million. When client property budgets of $41,000,000 are also added, the PH group collected, disbursed, and accounted for $83 million in 1995.

The original single company with everyone doing everything evolved into a group of three companies, each with its own clearly defined role in community building (Table 4.2). There is the parent company, Phipps Houses, which is owner, developer, coordinator, and final decision maker. Phipps Houses Services, Inc., is the property manager, both commercial

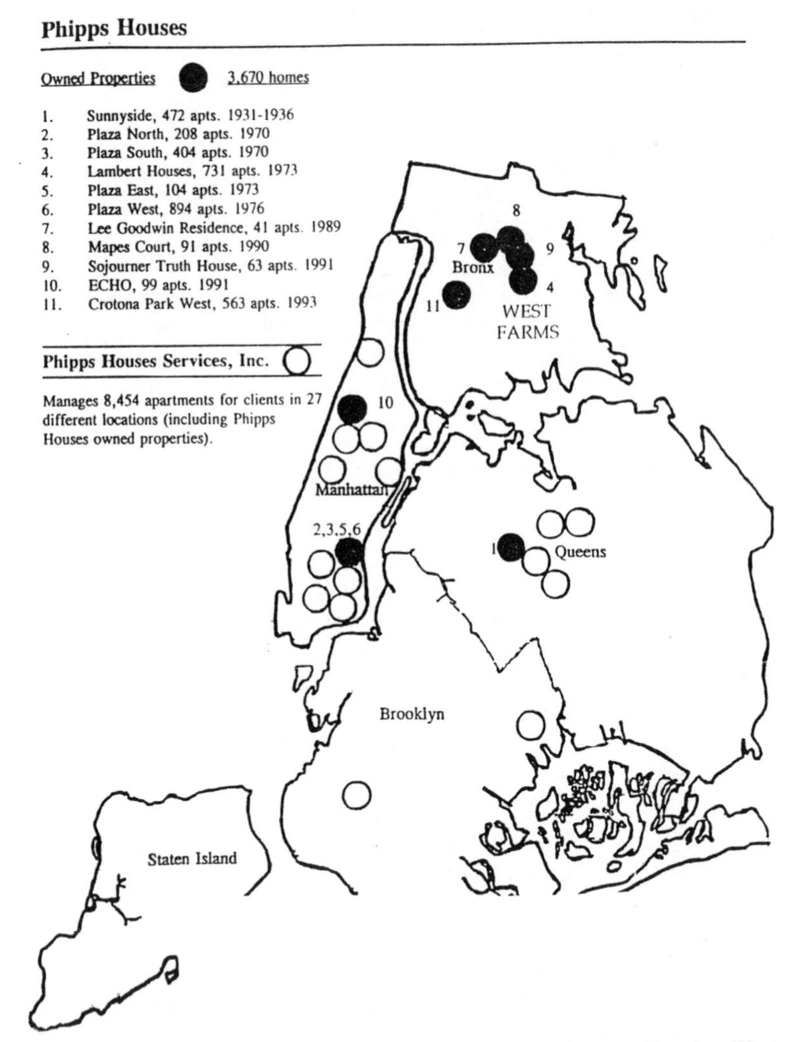

Figure 4.1. Map of New York City and Phipps Houses' Presence, Showing West Farms in the Bronx

and residential, and has by far the largest staff (to manage the 8,500 apartments in its 1995 portfolio). The Phipps Community Development Corporation (PCDC) facilitates individual and family enrichment and the

TABLE 4.2 Organizational Structure and Functions of the Phipps Houses Group, 1995

Entity	*Founded*	*Functions/Responsibility*	*1994 Staff*			*1995 Expense Budget*[a] *(in $)*
			Full-Time	*Part-Time*	*Total*	
Phipps Houses	1905	Developer and owner of the real estate in the Bronx, Manhattan, and Queens. Parent of affiliate and subsidy companies (PHSI, PCDC, and housing companies). Provider of technical assistance to not-for-profit organizations	15	0	15	$ 2,286,000[b]
		Sunnyside Gardens, Queens (472 apartments)	13	0	13	2,566,000
PHSI	1969	Property manager for all Phipps Houses' properties (3,670 apartments) and properties owned by other not-for-profit institutions, including several hospitals (an additional 4,851 units).	400	75	475	4,017,000
PCDC	1972	Facilitator of social communities within the buildings. Provides services to homeless, low- and moderate-income residents. Coordinates relations with surrounding areas and services by outside providers. (Budget includes CDD division of Phipps Houses.)	35	75	110	4,652,000
Total			463	150	613	$13,521,000

NOTE: CDD = Community Development Department; PCDC = Phipps Community Development Corp.; PHSI = Phipps Houses Services, Inc.
a. Net of subtenant income, depreciation, borrowing, and capital expenditures.
b. Not including the properties owned by Phipps Houses, each of which is a separate corporation (except Sunnyside, shown here).

development of social structures in the buildings. The group has shaped its internal and external communications systems to help ensure feedback among all the companies, and from residents and others.

Genesis of a New Philosophy: Enduring Communities Versus "Housing"

Within a few years, it became clear to PH that the provision of housing alone could not accomplish its goals: to help create better lives through beautiful and supportive communities—which alone can be enduring communities. People, not buildings alone, are the key to better lives. To PH, better lives meant stable personalities, characters, family life, and education such that individuals could be economically self-sufficient and fulfill their own purposes in life. This was an explicit rejection of the assumption of traditional housing programs, that people to whom the housing was targeted were held back primarily by bad physical living conditions—that they possessed the skills, habits, and attitudes to do well in life if only the obstacle of deteriorated homes was removed. But for PH, creating the buildings was only the beginning.

New York City's economy was losing its manufacturing base, and low-skill jobs were rapidly diminishing. Many of its new residents needed much more than a decent apartment. They needed education, recreation, friendship, mentoring, and a view of the world in which they were wanted, were capable, and could make fulfilling and economically independent lives. The basic premise of PH in 1974 was, and remains today, that the best place to reach people enmeshed in a cycle of poverty is their home place, making available in the apartment complex itself a program of skills enhancement carried out with personal caring.

During this time, PH also strengthened its commitment to good physical design, attentive to function, maximum sunlight, ventilation, proportion, texture, and color as well as economy. Although the physical environment cannot create community, it can be a powerful obstacle to or facilitator of efforts to form community. If there is no physical space where people can get together in an apartment complex, their difficulties in functioning as a community multiply. In poor communities, this is especially important. For example, community meeting and recreation rooms are essential to developing the connections between people that are the substance of community.[3]

■ The Early Years, 1969-1981: Working Within Instability

PH's first construction project in West Farms was Lambert Houses. With 731 apartment homes and around 3,000 legal occupants (doubling-up of families added hundreds to that count), as well as a shopping center, it is the size of many American small towns. If successful, it would be a model that could guide similar efforts elsewhere. The tenor of this early period was set in 1971 when building began without efforts by the contractor to hire local labor. Snipers on nearby roofs shot at workers and knife-wielders slashed car tires. The violence stopped when the contractor agreed with the "coalition" leader to hire local workers also.

Lambert Houses was completed in 1973. The new community began with difficulties. Within a few years, hidden poor construction and design of parts of the buildings became apparent. This and other defects resulted in lawsuits filed by Lambert to recover money for permanent repairs. The suits, won after 14 years of legal obstructions by the brick suppliers and contractors, siphoned money and management time away from maintenance and other constructive activities.

Despite the construction defects, large, comfortable apartments and pleasant grounds gave residents excellent living conditions unavailable to low-income people elsewhere. Forty-seven percent of the units had three bedrooms and 6.5% had four bedrooms, a ratio of large apartments not seen before or since. So many homes for large families met a huge need, but also meant the presence of about 2,000 children and youths. This translated into hard wear and tear and high maintenance costs and gave urgency to recreational programs, begun in 1974.

The first oil crisis of 1973-1974 tripled Lambert's heating cost. This necessitated rent increases beyond the ability of most residents to pay. Those whose incomes were high enough had other housing options, and gradually many moved away. They were replaced by poorer people, many on welfare, supported by Section 8 subsidies.

West Farms is part of the larger Tremont area. PH established the Tremont Improvement Program (TIP) in 1974 to assist with recreation, job services, and housing. PH assisted TIP with money and management oversight until TIP's independence in 1981. The Nixon moratorium on housing development doomed the housing part of this initiative. Later, TIP became the TIP Neighborhood House (TIPNH), which raised funds for and ran social, educational, and recreational programs. Its federally funded talent search program sent around 700 young Bronxites to college and received presidential recognition. Financial mismanagement led to TIP's demise in 1991.

The 1970s were a decade of destruction and instability. Rising costs eliminated landlords' profits. With no prospect of higher rents, there was widespread abandonment. Scenes were televised across America, showing Bronx neighborhoods looking like Germany after World War II. Landlord arson (to claim insurance payments) led to further destruction. Between 1970 and 1980, the Bronx lost hundreds of buildings and over 300,000 people. In West Farms, major sections were abandoned (Figures 4.2 and 4.3). Its population dropped from 21,000 to 13,000, a loss of nearly 40%.

By the mid-1970s, PH did a survey of the neighborhood, seeking to forecast the progress of abandonment and to propose actions to stem it. It created a plan for physical renewal and for founding an organization to train and involve residents in actions for the area's betterment. The city and foundations rejected it because of the substantial funds required to fund it and their bad experience with Great Society antipoverty corporations. It would be nearly 20 years before comprehensive renewal efforts return to favor. PH also set up the Bronx Park South Management Corporation (BPSMC) for a trial run of better property management. However, the problems far outran resources, just as they had for private landlords.

This period saw another population shift: Clear Black dominance ended with the creation of a strong Hispanic majority as Spanish speakers moved into newly renovated Section 8 units and as new immigrants took existing vacated apartments, however rundown or uncertain the tenure. The consequences to West Farms' social fabric were significant. Transformation requires continuity; there was none in the West Farms of the 1970s.

A successful initiative was the Bronx 2000 Local Economic Development Corporation, founded in 1979. To consider the area's future, PH organized a conference in 1978, which recommended economic development. The board of Bronx 2000 soon included a majority of local residents and institutions' staff. Its period of financial transfusion and management tutelage under PH was relatively short.[4] Bronx 2000 did commercial storefront improvements, ran a Farmer's Market, was administrator for NYC on numerous foreclosed abandoned buildings, and still works extensively with the Bronx Housing Court. It also owns a large recycling facility that provides income to local residents and produced an important study showing the influence of drug activity on the loss of housing.

At the time, the Bronx was plagued by a long history of dishonest political leaders who dispensed public jobs, building projects, and patronage to those who paid bribes or otherwise returned favors. The boss of the Bronx, Democratic County Leader Patrick Cunningham, resigned under fire for corruption in 1978 and was convicted and sent to jail a few years later. His successor was Stanley Friedman, whose empire of corruption dominated every corner of the Bronx of the 1980s. He and his protégé, Borough President Stanley Simon, were exposed in 1986, and Friedman, along with former Congressman Mario Biaggi, went to jail in the late 1980s. Simon was indicted and convicted of bribery and extortion.[5]

In this environment, Phipps Houses received no development projects until after exposure of the corruption in 1985. Then it was awarded only projects that for-profit developers did not want—homes for the very poor and homeless, yielding no profit but requiring great knowledge and effort.

West Farms Pushed to the Bottom

During the early 1980s, the empty hulks of West Farms' many abandoned buildings were occupied by drug dealers and users. Just walking to the grocery store was dangerous because of flying bullets and robbery by addicts desperate to support their habits. When New York City finally demolished the buildings in the mid-1980s, about one third of West Farms' land area was vacant.

This was a decade of subtle but devastating governmental destruction of the social fabric in West Farms. Real incomes of poor people dropped significantly. This, combined with the federal withdrawal of support from social housing and welfare programs and the rising drug pervasiveness, increased conditions of hardship in West Farms. The U.S. Department of Housing and Urban Development (HUD) had been setting its contribution through Section 8 such that a family paid a maximum of 25% of its income for rent. In 1981, legislative changes increased this maximum to 30%.

HUD's blockbuster change, however, was in its rules for eligibility for Section 8. The maximum eligible income had been 80% of area median income, allowing a majority of the working poor to qualify. Reserving subsidy for "the truly needy," HUD cut the maximum income to 50% of median. This, combined with federal preferences,[6] meant that most vacant apartments had to be rented to (new) residents who were very poor, usually

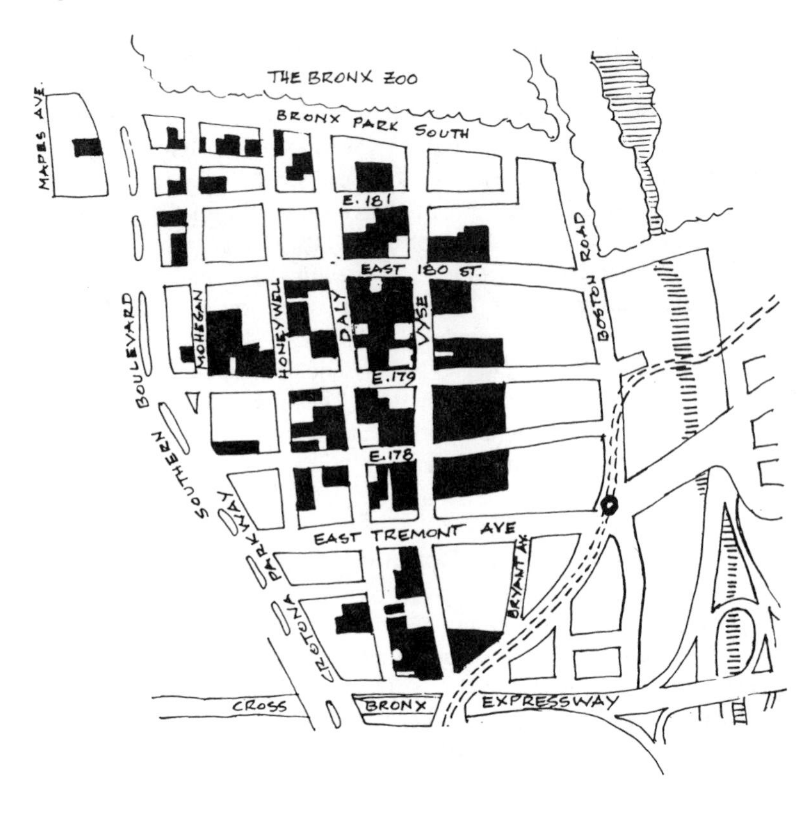

Figure 4.2. City-Owned Vacant Land, 1990: Bronx Park South Urban Renewal Area

on welfare and practitioners of the welfare ethic dictated by the system's eligibility rules: You can't work or get married without losing benefits. In 1994, it would have taken a job paying nearly $25,000 to replace welfare (including Medicaid and food stamps), the Section 8 subsidy, and the earned-income tax credit benefits for a family of three.[7] Few recipients qualify for such jobs, and taking a lower-paid job, resulting in a loss of Medicaid and less disposable income, is not a rational choice for most.

The deleterious effects of HUD's policies of the 1980s on West Farms cannot be overestimated. The new residents were less educated, less

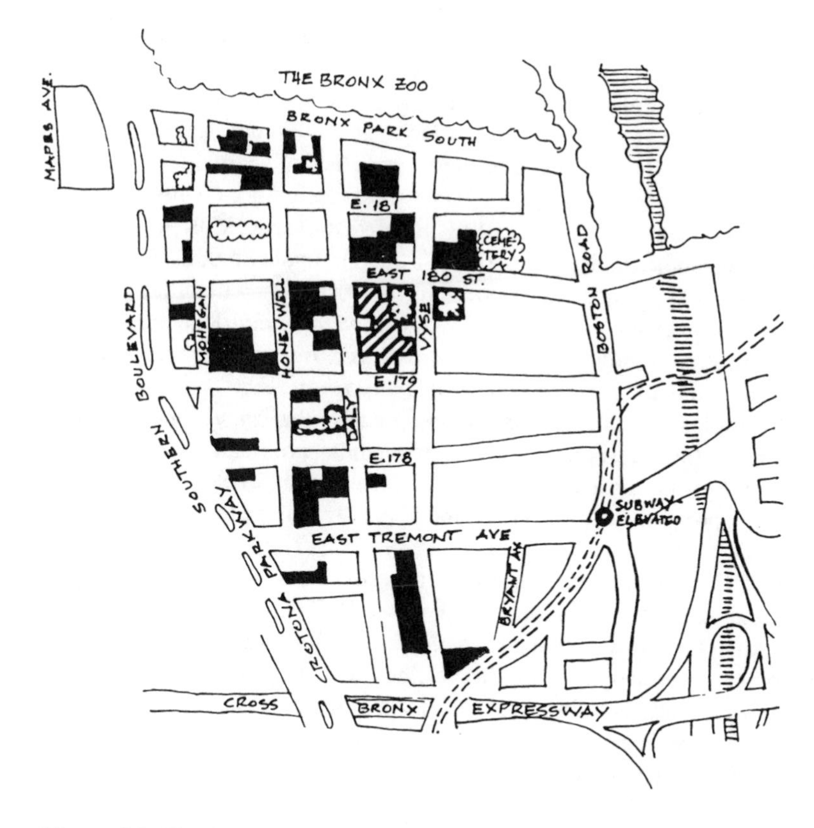

Figure 4.3. City-Owned Vacant Land, 1994, plus Gardens and Open Space

stable, and less able to fend for themselves in society. Many Lambert children never saw an adult get up in the morning and go to work. Lambert was home to fewer and fewer good role models and subject to more and more vandalism. The drug trade increased everywhere with its concomitant increases of street violence and abuse of women and children. The public environment—streets, sidewalks, vacant lots, and parks—became unusable for the play and social activities that build relationships. Traveling from home to school or grocery store could be an event of mortal danger. Outdoor public life was virtually destroyed. Any civic life struggling to emerge had to take place indoors.

Phipps Houses' Response: Plans for Regeneration

In the early 1990s, West Farms residents knew there was no place else for them to go. They were, by and large, at the bottom of the socioeconomic ladder. They tended to be young and Hispanic, with limited education and low incomes. Sixteen percent were foreign born. Only 43% had graduated from high school; 49% lived in poverty. Most households reported incomes under $10,000, and a quarter, under $5,000. Unemployment was very high. Sixty-six percent of households with children were headed by women. Hundreds of families were living doubled up. The concentration of poverty in West Farms was extreme, the direct result of the uncoordinated policies of all levels of government.

PH initiated a multifaceted response to these challenges. In addition to various actions responding to the immediate problems, it decided to survey West Farms again and create a second plan, to carry out a major building program according to it, and to greatly enlarge the size and scope of PCDC.

PCDC had been founded in 1972 to receive tax-exempt donations for PH community activities. A strengthened PCDC was seen to be critical in attacking the growing social problems through citizen education and involvement in neighborhood issues, and in founding and coordinating local organizations. It involved reconceptualizing the scale of operations due to the residents' increasing needs; refinements to its tax status; a search for leadership at a higher level; and the development of a strong professional fund-raising capacity.

Social changes in West Farms were built slowly until the more visible, physical changes helped bring about a new enthusiasm. The exposure of the city's pervasive corruption and the victory of David Dinkins, New York City's first Black mayor, helped remove obstacles to PH receiving governmental resources. In 1988, a new, young Hispanic borough president of the Bronx, Fernando Ferrer, was elected; he had formerly represented West Farms in the city council and had progressive ideas. City and state governments had begun to replace some of the lost federal funding. By 1990, the election of President Clinton brought to power an administration committed to the sort of programs PH had laboriously put together privately during West Farms' "dark ages." Government and foundations began to see that the focused, comprehensive approach to neighborhood renewal could work to solve the complicated and interrelated social and physical problems of inner-city neighborhoods.

The Rebirth of Civic Life, 1990 Onward: "New Hope and New Housing for the Bronx"

The 1990s are seeing the fruition of PH's 25 years of work in West Farms. The new climate of political acceptance of PH's focused, comprehensive neighborhood renewal approach and its growing reputation based on its accomplishments, during 20 years in the Bronx, have led to an explosive growth in resources. The cumulative effects of past work have begun to pay off (Table 4.3).

Physical Renewal

In 1990, PH was involved in the planning and renovating of four new communities located in West Farms. Others joined, continued, and expanded these efforts, called "New Hope and New Housing for the Bronx." Two of the new developments were for homeless mothers and children. In them, PH sought to attend to the human as well as the shelter needs of the residents. Funded by the city and state, the residences were to shelter mostly young families for up to 12 months while they actively prepared for independent life. Education and social services were an integral part of the housing program. PH developed two such residences, Lee Goodwin Residence (1990, 39 apartments) and Sojourner Truth House (1992, 63 units), each containing permanent as well as transitional apartments. PH was joined by Women in Need, an organization specializing in services to mothers and children and in alcoholism work. Its role is to deliver counseling in education, employment, and housing, as well as day care to allow mothers to attend school and find housing.

In Rivercourt Homes, PH worked to bring in higher-income homeowners to join the mostly Hispanic owners of one- and two-family houses who had stuck it out through the devastation. It was PH's first venture into manufactured housing and its first home-ownership community, located adjacent to Lambert Houses. The goal was twofold: first, to provide large, semiprivate dwellings (each of the 33 homes contains a 1,500-square-foot owner's apartment with front and back yards and a rental apartment) to which West Farms residents might aspire when they did better in life (instead of having to leave the community) and, second, to introduce into the area people with steady jobs, stable lifestyles, and a vested interest in the well-being of their surroundings. Each house received a $50,000 state subsidy and was sold for around $156,000 within 6 months of completion

TABLE 4.3 Phipps Houses' Residential Real Estate Development Periods, 1905-1995

Year	*Type*	*Funding*	*No. of Units*	*Where*	*Residents (mainly)*	*Goal*
1905-1906	Apartments	Self-funded	600±	Manhattan: East Side, West Side	Foreign and in-migrants (poor White and Black)	Decent housing for working people
1933	Apartment community	Self-funded	472	Queens: Sunnyside	Working class (moderate-income White)	Decent housing for working people
1955-1965	Apartments	Client funded	750±	Manhattan: Upper East Side	Hospital staff (moderate and fixed incomes)	Develop staff housing for hospitals
1965-1977	Apartment communities, neighborhood house	State, federal, and New York City subsidies and financing; tax shelters; Phipps Houses' equity	2,341	Manhattan, Bellevue, NYU Medical Center, Bronx	Hospital staff, Working poor (White, Black, Hispanic)	Urban and community renewal
1977-1995	Apartment communities (6); organizations: Bronx 2000 Local Econ. Development Corp., PCDC	Mixed government subsidies and grants; tax credits	1,489	South Bronx (1969-present)	Low and very low income; homeless (Black; Hispanic; many immigrants—Central and South American)	Focused coordinated human and community renewal in West Farms, South Bronx

NOTE: PCDC = Phipps Community Development Corp.

in late 1991 to families with incomes of between $32,000 and $53,000, half of whom came from outside the Bronx. The grounds of all the homes have been extensively improved with ornate wrought iron fences, window guards, flower beds, and sometimes elaborate terrace structures. Further success will be indicated if and when Rivercourt's owners turn their attention to undertaking committee work and activities aimed at neighborhood betterment. In early 1996, there are signs that this may be happening.

The fourth West Farms development in the "New Hope and New Housing" program is Mapes Court, a two-building renovation that opened in mid-1990, with 91 permanent, large rental apartments. It was developed with the federal tax credit incentive and is therefore limited to a maximum tenant income of 60% of median income, with a 10% set-aside for homeless families. Mapes proved to be an effective combination and balance of size, program elements, and staff. Its small size is a major factor in its success. It has a garden and play space on the adjacent lot (vacated by a demolished building), as well as four community rooms opening onto the various grades outside the buildings. In the first 4 years, two full-time young women workers, Black and Hispanic, were excellent role models for residents. At least 50% of households attend monthly Resident Association meetings and many are pursuing better lives through obtaining their GEDs and enrollment in college and training programs. Getting off welfare has gained currency as a good thing. The seniors' garden produces food for families out of money at the end of the month as well as for the gardeners.

The final part of the PH physical program is the open space network. Where there was not one foot of public open space, residents are turning vacant lots into lush green oases. They have the assistance of the city's Green Thumb program and the Bronx Botanical Garden, which provide fences, seeds, tools, and instruction. The 1990s saw five new gardens and two new green open play spaces with gardens. To bring much needed active play space, the borough president allocated funds for a new city park of 1.77 acres on the most central block. The open space network has become so central to civic life that it is unlikely that residents will allow the city to put housing on all of its vacant sites.

Social Renewal

In 1990-1991, the strengthening of PCDC began to bear fruit. PH commissioned a study of children's play patterns,[8] convening interested parties (parents, teenagers, teachers, service workers, and citizens) in pursuit of a practical neighborhood issue. This active participation of a wide mix of residents had not been seen in many years. Concurrently, PCDC also organized the West Farms Consortium of directors of local organizations, mostly housing and social services, to deal with matters of shared interest. It began with the coordination of services for client families in common and broadened to include other issues affecting the area, transmuting into a Neighborhood Advisory Council.

Within Lambert, PH expanded the programs for residents. Reaching as many individuals as possible is important not only for the effects on individuals but also for the cumulative numbers of people constructively involved. "Many yeasts leavening many breads" increases the critical dynamic mass, which can change a community's norms. Personal contact is a vehicle for personal and social change. It can lessen the learned helplessness acquired by many growing up in an unremittingly harsh environment. Lambert women now form the nucleus in various advisory councils, determined to take back turf from the drug business and to enlarge the sphere of constructive life in West Farms.

PCDC increased its reach to individuals through its Family Advocates, social workers available for people suffering from domestic violence, depression, or difficulties with welfare administration, children's education, or health. It is supported on a permanent basis. The Home Instruction Program for Pre-School Youngsters (HIPPY) is a home-based educational effort in which paraprofessionals teach mothers how to be the first teacher of their 4- or 5-year-old child. They use a highly structured package of materials to teach basic life skills such as arithmetic, reading, and self-discipline. Its multiplier effect is remarkable (uncles, aunts, siblings, friends) and its effectiveness, exceptional. A federal Head Start grant funded the institutionalization of HIPPY in Lambert in 1995.

The Teen Parenting program reached girls and boys, 14 to 22 years old, teaching parenting skills, counseling for employment and education, and family planning. A Teen Outreach van travels the streets of West Farms seeking young people who are just "hanging." Workers talk with them about facilities available for sports, recreation, and counseling. The hope is to attract them into any program, which then becomes their entry into a coordinated individual program directed to a better future.

A major development was the Comprehensive Community Renewal Program (CCRP), created by a consortium of foundations to assist well-established organizations in the South Bronx develop further in the direction pioneered by PH in the 1970s. Substantial 3-year grants were a significant validation of PH's approach. CCRP brought a new and higher level of funding and synergetic networking among leaders, research, and management development to its members. It financed preparation of a Quality of Life Plan in which urban planners John Shapiro and associates acted as facilitators for community meetings called to highlight the worst problems and to elicit residents' ideas for change. PH took the lead in promoting and creating this component of CCRP, which in 1996 won the main award of the American Planning Association. Published large-scale plans now embody local leaders' multi-year agendas. CCRP funds have allowed PCDC to better coordinate its many programs as well as install computer equipment for management and tracking of data on programs and outcomes.

The expansion of PCDC's work has recently included initiatives in health care, employment, and use of a local school as a community center. Mapes Court houses the Bronx Perinatal Clinic and Lambert's shopping center replaced a departing bank with a Family Health Practice. Youth Fair Chance partners PCDC with the South Bronx Overall Development Corporation in an 18-month employment program for youths, both in school and out of school, with $3.2 million funding from the U.S. Department of Labor. The Beacon School is a city program, funding the use of selected public schools after school hours and on weekends. There are 5,000 attendees every month at PCDC's Beacon, drawing the entire West Farms community. The concepts of family service and case management undergird the many programs.

These PCDC programs expand the founding idea of "reaching people where they live" by adding the idea of "connecting home to neighborhood." Young people and adults can start in a protected atmosphere before going out into the larger world. This reaching out eventually connects West Farms to the rest of the Bronx and New York City.

■ Outcomes and Lessons

For West Farms

1. The neighborhood is well on its way to physical reconstruction. In the past 25-plus years, nearly 2,700 apartments have been built or rehabilitated. Together, they have added around 12,500 people to the neighborhood in a variety of tenure types. This is virtually all of West Farms'

present population. In addition, an extensive new open space network is being created. It will provide the "outdoor living rooms" in which community activities can take place.

2. In 1996, there is a vital civic life emerging, as evidenced by new local leaders, organizations, and institutions dedicated to West Farms' betterment, many begun by PH. Government agencies are now familiar with West Farms as a neighborhood on the move; they have seen it as a good place to fund programs because its accomplishments to date give new programs a better foundation and chance for success. PCDC itself developed as an institution to facilitate West Farms improvements, and it will continue to operate.

3. Constructive thoughts, habits, skills, attitudes, and hopes have emerged and are asserting themselves. They reinforce, for example, the notion that education is to be valued and is a vehicle for social and economic advancement; that work can be gratifying and is an expected adult responsibility; that girls and boys, men and women are of equal worth, intelligence, and importance; that self-expression is an essential experience for children; that domestic violence is not acceptable behavior; that a person can affect the course of his or her life; that health care is available through clinics and promoted by better nutrition; and that some institutions will respond to the residents' needs.

4. In recent years, thousands of West Farms residents have had personal contacts and relationships with good role models: women and men; Blacks, Hispanics, and Whites; city planners, teachers, counselors, board members, volunteer mentors, and technical staff. This has dispelled many racial stereotypes on all sides.

For Phipps Houses

Working in West Farms for 25 years has transformed PH organizationally and its staff personally.

1. Knowledge of changes and intensifying needs in the South Bronx led to expansion of both physical and social development programs.

2. Over many years, PH developed a successful fund-raising program that greatly enlarged its potential for action.

3. The community development function of PCDC grew from its original, minuscule size to that of a full partner in PH, with responsibility for development, ownership, and property management.

4. Where appropriate, PH companies function as employers-trainers for residents.

5. PH has changed from a small, centrally directed, traditional, single-focus housing company to a large, integrated, full-service real estate operation, which involves residents.

For the City and the Country

1. Large-scale application of focused, comprehensive human and community renewal will require public education that this kind of neighborhood and human renewal works. The biggest costs come from not doing it: drugs, crime, prisons, and subsidies of all kinds required to feed, clothe, and shelter dysfunctional, and often depressed, people. Financially, the task may be expressed as investing money to convert a huge productive potential (see Table 4.4 for PH's costs).

2. A wide array of programs is needed, each with specific goals and continuous feedback to measure success. Strong incentives and disincentives are needed to motivate some noncriminal residents toward positive transformation. Help close at hand, in the home place, will be critical.

3. To renew a neighborhood's physical and social structures is a large undertaking. It is accomplished in day-to-day increments. A long-term view (20 years or more) is essential. Continuity is the key to success. People must know that institutions will be there over many years and that they can rely on them.

4. Staff continuity and professionalization are equally important. Most not-for-profit housing groups can pay only the meager salaries dictated by fees and grants from government agencies and foundations, which are based partly on lack of knowledge of real estate and consequent blindness to the high cost of incompetence.[9]

American society has entrusted not-for-profit housing organizations with stewardship of billions of dollars' worth of real properties, paid for by taxpayers, sheltering hundreds of thousands of people. Social housing often suffers from inadequate reserves and inadequate construction and is often left to poorly paid, insufficiently qualified staff.[10] If taxpayers' investment and society's assets are to be preserved, foundations, government, and business must set about professionalizing social housing. It will be by far the least expensive course.

5. Privatization of social housing in the absence of changes to raise the maximum incomes permitted at entry could lead to further, faster decline. Making profits on most buildings can be done only by forgoing

TABLE 4.4 Phipps Houses' West Farms Residential Communities, 1973-1995

Entity	Total Development Cost (in $)	Year Completed	No. of Households	No. of Residents (approx.)	% Welfare (approx.)	Ratio[a]: 1995 Staff	Ratio[a]: Full-Time Staff per Household	Ratio[a]: Staff Person	Estimated 1995 Costs (in $) per[a]: Household	Estimated 1995 Costs (in $) per[a]: Person	PHSI Staff: Property Mgt. & Maintenance
Lambert Houses	32,000,000	1973	731	2,842[g]	41	24 FTE[e]	1/30	1/126 }	4,337[i]	1,172[i]	4.0 + 31
Mapes Court	9,300,000	1990	91	242	36	1.5 FTE[f]	1/61	1/161 }			0.8 + 3
Lee Goodwin Residence[b]	4,900,000	1989	40[d]								0.7 + 2
Permanent			(12)	46	100	0.6[g]	1/20	1/77	1,400[j]	365[j]	
Transitional			(28)	98[h]		18.0	1/5	1/5.5	17,000	4,850	
Sojourner Truth House[b]	7,000,000	1991	62[d]		90						0.3 + 3
Permanent			(29)	77		0.4	1/72	1/192		385	
Transitional			(33)	87[h]		20.0	1/1.65	1/4.35	14,850	5,630[j]	
Rivercourt Homes[c]	6,700,000	1991	66	208	0						
Total West Farms	**59,900,000**		**990**	**3,600**							**5.8 + 39**

NOTE: PHSI = Phipps Houses Services, Inc.; CDD = Community Development Department.
a. Phipps Community Development Corp.-CDD data.
b. Residences for homeless mothers and children.
c. Consists of 33 two-family private homes.
d. Excluding superintendents' apartments.
e. Had a full-time staff of 16 plus 20 volunteers counted as 8 full-time equivalents (FTE) in 1994.
f. Counts the property manager of PHSI as 0.5 FTE in 1994. In 1995, CDD staff dropped to 0.6, or 1/403 persons.
g. Family housing occupancy: Studio, 1.0 person; one bedroom, 2.0 persons; two bedroom, 3.75 persons (Rivercourt: 2.0 persons); three bedroom, 5.0 persons (Rivercourt: 4.0 persons); four bedroom, 7.0 persons.
h. Homeless housing occupancy: Studio, 2.0 persons; one bedroom, 2.5 persons; two bedroom, 4.00 persons; three bedroom, 5.0 persons; four bedroom, 7.0 persons.
i. Costs are higher because of large units and families in Lambert Houses and more programs delivered, and because most programs serve all of West Farms in addition to Lambert Houses and Mapes Court.
j. Lee Goodwin Residence's and Sojourner Truth House's service level here is determined by New York City, New York State, and the federal government.

maintenance. This is so because the only other immediate source of income is increased rents, paid either by tenants or government. But governments are not likely now to raise subsidies to pay private profits. Nor are tenants, many of whom are as yet ill educated for the workforce, likely to increase their incomes soon. Indeed, as welfare reform proceeds, many more residents of social housing will be scrambling to make ends meet. Alternatively, raising income limits to admit working families would promote the idea that work is a good thing as well as provide a source of funds to replace diminishing governmental subsidies.

6. PH has shown that focused and comprehensive neighborhood renewal works. This approach appears to be the best hope for ameliorating the ills of the inner cities. PH provides a model for an organization structured to produce fiscal responsibility and broad-ranging action. It involves people in their own transformation by engaging them in solving the problems they face, including drugs, inadequate housing, crime, inferior schools, and lack of medical services. It is a relatively slow process and requires long-term commitments. Nevertheless, instituting (nationwide) comprehensive, focused, local community renewal can "reweave" damaged social fabrics and give individuals and families a context in which self-change makes sense.

NOTES

1. Robert Caro's (1975) biography of Robert Moses devotes two chapters to the larger Tremont area, its organized fight against the Cross Bronx, and the effects of the process on West Farms.

2. Quote is from land disposition agreement between City of New York and Lambert Redevelopment Company, December 3, 1970, certifying Lambert's eligibility for urban renewal funding.

3. HUD's regulations did not allow community rooms; PH called them "storage" on the plans to have them approved.

4. In supporting the Bronx 2000 director's right to choose his own staff, PH incurred the enmity of local political powers intent on keeping control of everything. PH was blackballed for certain funding sources for years afterward.

5. The Bronx was not alone. There was equally serious corruption of leaders in the boroughs of Queens and Brooklyn. See Newfield and Barrett (1988) for a contemporary, detailed, journalistic account of these years.

6. Mandated federal occupancy preferences were for the homeless, for those paying over 50% of income for rent, and for those involuntarily displaced from their housing.

7. See Fox (1994). John L. Fox is a trustee of PH and former director and officer of the American Institute of Certified Public Accountants.

8. From the Center for Children's Environments of the CUNY Graduate Center, completed in 1992.

9. TIP's original director went to work for a major American television network's news division. She became the London-based head of finances for Europe, Africa, and the Middle East. She said it was much easier than running TIP.

10. See Bratt, Keyes, Schwartz, and Vidal (1994) for an excellent and detailed national survey that states the problems well.

REFERENCES

Bratt, R. G., Keyes, L. C., Schwartz, A., & Vidal, A. C. (1994). *Confronting the management challenge: Affordable housing in the nonprofit sector.* New York: Community Development Research Center, Graduate School of Management and Urban Policy, New School for Social Research.

Caro, R. A. (1975). *The power broker: Robert Moses and the fall of New York.* New York: Random House.

Fox, J. (1994, August). *Public assistance in New York City.* (Unpublished study available from Phipps Houses, 43 West 23rd Street, New York, NY 10010, Attn: President's Secretary)

Newfield, J., & Barrett, W. (Eds.). (1988). *City for sale.* New York: HarperCollins.

West Farms Planning Task Force. (1994). *A quality-of-life physical plan for West Farms in the South Bronx.* New York: Phipps Community Development Corporation and the Comprehensive Community Revitalization Program.

Part II

Turning Around Failure

PETER MARCUSE

Housing is always at once a most private and a most public matter. For the user, the home is his or her castle, protected from outside invasion in the Constitution, a symbol and expression of individuality, a constituent of the self-image and external representation of the person that lives in it. "Private" has a further meaning for builders, owners, managers, and others in the private real estate industry: the opportunity, on their own initiative, without state compulsion, to carry on activities relating to the supply of housing in order to make a private profit (or in the case of nonprofits, to achieve goals that the private organization freely sets for itself).

But housing is not only private; it is at the same time inextricably linked to public action: Providing and living in housing in an urbanized society are dependent on collective action, inter-relationships, contacts, and intensive and ongoing government activity. Without roads, sewers, a water supply, a police force and a fire department, building codes, zoning regulations, and laws and courts to enforce them, no civilized housing provision can take place.

The three case studies in this section revolve around the search for the "right" relationship between these private and public interests in housing. Their successes (and their limitations) are related to the way in which public entities are willing to deal with private ones. Three general lessons stand out:

- The more intense the direct involvement of users in the control of their housing conditions, the more likely is their ultimate satisfaction with that housing. Control means that, at a minimum, users have decision-making power over housing provision, maintenance, and management, but not necessarily that they undertake such work themselves. *Userization* or *communitization* are possible terms for this.
- The more flexible and responsive to users the public role in the provision of housing can be, the more satisfactory and the higher the quality of the outcome is likely to be. This often includes the involvement of private business entities (for profit or not) in the relationships between public authorities and private users. *Privatization* is the conventional term here.
- Housing is an expensive proposition, and quality results require quality and quantity investment that for poor people means substantial and ongoing subsidies. Scrimping here can undercut all the benefits of implementing the first two principles. *Subsidization* is what is required here.

Thus, the three recommendations that can be drawn from these case studies concern *userization, privatization, and subsidization* in combination, not separately.

User control can be at three levels: broad policy, direct decisions affecting development and management, and the direct undertaking of detailed day-to-day work. The first is probably the rarest. Only the New York City example provides a model in which residents and their organizations fought with their private landlords, then with the city as landlord, to shape a set of policies that would give them control over their own housing. They organized and agitated, pushing not only the immediate housing bureaucracy but also ultimately the mayor himself to support a program that was expensive for the city but was seen by the political leadership as having a payoff in votes. In the process, they developed a relationship of mutual respect with the city's housing bureaucracy and became a consistent participant in the policy-making process at the citywide level.

In Boston, the political role of tenants was likewise essential, although developed at the neighborhood rather than citywide level. Their participation in the Monastery Hill Planning Task Force was active, informed, and technically competent but backed by grass-roots organization. It ultimately resulted in their designation as the "best risk" for heavy investment in upgrading and gave them, through cooperation with other influential forces on the Task Force, the clout to carry through the principle of user decision making from the planning through the implementation down to the continuing maintenance of their housing.

Direct authority over development and management was central in both New York and Boston to ensure the improvement of housing conditions. The daily monitoring of rehabilitation in New York and the detailed negotiation of specifications for the work to be done in Boston were conceded by all parties to be central to the respective work. No wonder; who would have a greater interest in making sure that contractors performed properly than those who, in the end, would use the results of the contractors' work? In Boston, professional help was hired to ensure competent work; in New York, the tenants themselves did the supervision but at substantial emotional cost to themselves. But they did it, and the quality of the product attested to their involvement.

Ongoing day-to-day involvement, whether in construction or in management, is another story. In New York, budgeting, the central control point in management, was ultimately surrendered by tenants to professionals hired to do it, and outside technical assistance played a key role. In Boston, paid staff and hired independent professionals were used in review and supervision processes. The possibility of a Tenant Management Corporation, with tenants engaged full-time in management, is still under discussion. The major incentive is the possibility of increased subsidies rather than the desire of tenants to get involved in the actual performance of managerial or maintenance tasks. Again, that seems logical enough: housing is not the career of choice for most people. On a paid basis and certainly on an unpaid basis, it is hard, often thankless work that most residents would prefer not to have to do for themselves, unless there is no other alternative.

The Chicago experiment, whose results still seem very much in doubt, may in fact have the lack of resident involvement at either the policy or the implementation level as one of its weaknesses. If positive relations among residents (particularly those with low to moderate income) were a primary concern of the MINCS program, then their early involvement in structuring the program might have promoted greater interaction than was found to have taken place. Hiring tenants to do direct maintenance works to their own advantage and might be very desirable in improving the quality of work, although the evidence thus far is inconclusive as to whether choosing workers because they are residents is more effective than choosing them for their skills. A combination is obviously ideal, but not always possible.

Privatization played a role in only one of the three cases: the City of Boston.[1] What it accomplished was to create a simple and clear way of providing accountability to the residents for work done by those doing it—if the residents were dissatisfied, they could fire the contractor. Public entities might quite possibly work as efficiently as private ones, but certainly the ability to control them is more complicated than what the Boston private contract arrangement provides.

Finally, subsidization: the issue of money. The figures given in these three cases are startling: up to $60,000 a unit in New York, $100,000 in Boston, and $70,000 in Chicago. These approximate new construction costs, but they were for the rehabilitation of existing units, not to meet any luxury standards. With all the attention focused on each of the three projects, holding down costs would have been an ongoing concern; yet, no lesser investment could produce the results desired. The lesson here is not a popular one: good housing costs money—substantial money. The fond hope that some hitherto unexplored housing "innovation" would suddenly show how better housing could be provided and money could be saved at the same time does not receive support from these cases. Self-help, in the sense of userization, makes major contributions to improved housing; but its advantage is not in saving money, at least not in the short run. It is deception, perhaps self-deception, to believe otherwise.

These three cases, then, show what can be done by combining tenant leadership with public support for and confidence in the ability and concern of tenants to maintain and improve their own housing. Two cautions, however, need to be expressed.

First, showpiece projects are not likely to be easily replicated, particularly on a large scale and even more so when they are expensive. In New York City, the TIL program has only affected a minority of city-owned housing units; the more its scope is expanded, the more an investment is required in organizing and supporting tenants' own efforts. In Boston, the Commonwealth is the "best-risk" project; another, not so good-risk project was apparently not accounted a success. The MINCS program was not even taken up by any city but Chicago, its originator, again apparently because of the financial obligations for any authority wanting to undertake it. Hence, replicability depends on a substantial increase in public funding, an increase that would be highly desirable but is not very likely, at least not from federal sources today.

Second, each of the cases excludes "problem" tenants from their benefits. Tenants in New York need to steel themselves to evict nonpaying residents; in Boston, residents insist on careful screening of new applicants, and in Chicago, the Housing Authority itself screened them rigorously. Tenant selection and control of tenant behavior is unquestionably critical for good housing management, but its down side is the unmet needs of the tenants who are thus excluded. The supportive services needed to handle a broader range of tenants are what seems to be cut first in the reduced funding for the Commonwealth project; neither of the other cases suggests that they address the problem of the excluded.

These cases are also rich in other lessons, well discussed in the descriptions that follow: the importance and the limitations of dealing with basic systems—heat, hot water, sewerage, and so on—and of dealing with design improvements, "amenities," and flowers; the complexity of defining "success" in a housing experiment; the importance of neighborhood conditions to resident satisfaction with individual units; and the importance of dedication, imagination, and perseverance in dealing with housing problems. On other points, the evidence is inconclusive: the involvement of profit-motivated firms may be helpful but did not occur in two of the three cases. Income mixing may be helpful but was not involved in two of the three cases. All three cases come from large cities with political as well as financial resources; similar efforts may be less successful in smaller communities.

But one thing is clear from all three cases: attention to tenants, involvement of tenants, and the initiative and pressure of tenants are critical to successful housing for those today at the bottom of the housing-quality ladder. Good housing provision is not a question of real estate practices but of social and human policy.

NOTE

1. Some might call the TIL program's process of getting title to property out of the city's hands into those of the residents *privatization,* but that is a deceptive use of the term. There is a huge difference between a city transferring title to a private landlord who is running a building for profit, properly called privatization, and transferring it to the residents themselves for their own use, which I am calling *userization.*

5 The Revitalization of Boston's Commonwealth Public Housing Development

LAWRENCE J. VALE

Abstract

During the 1980s, the Commonwealth public housing development in Boston was transformed from a state of social and physical devastation into a nationally recognized model for public housing revitalization. This chapter examines the redevelopment process, emphasizing the role of the Commonwealth Tenants Association and its relationship with the Boston Housing Authority, the quality of site reprogramming and design, the innovative privatization of management functions, and the role of neighborhood groups and institutions in catalyzing the redevelopment effort. In so doing, the chapter raises broader issues about the nature of "success" in public housing redevelopment and examines implications of the Commonwealth experience for other "severely distressed" public housing projects.

This chapter examines the strengths and weaknesses of the comprehensive redevelopment effort undertaken during the 1980s at the Commonwealth public housing development in Boston, Massachusetts.[1] In the course of five years, the 648-unit housing development, known colloquially as "Fidelis Way" until the time of redevelopment,[2] was transformed from a state of social and physical devastation to a nationally recognized model for public housing revitalization. The dramatic physical reconfiguration of its low-rise and high-rise structures and the reprogramming of the site, combined with an innovative privatization of management functions, have made this transformation arguably the most celebrated effort to turn around the problems of a large public housing project anywhere in the United States.[3]

Drawing on recent interviews with a broad spectrum of residents about the process of redevelopment and current conditions, as well as on other documents and interviews with management, community organizers, and housing authority officials, the chapter examines both the successes and the limitations of the redevelopment effort from the points of view of those most involved in it or affected by it. The chapter examines the process by which the redevelopment came about, emphasizing the role of the Commonwealth Tenants Association, its relationship with the Boston Housing Authority, and its partnership with a private sector developer and management company. It describes and analyzes the redesign approach, one that employs the principles of "defensible space" in particularly innovative and effective ways throughout the development. It discusses the role of the surrounding neighborhood in the overall success of the redevelopment strategy. Finally, it considers the question of success—how it can be measured, explained, sustained, and replicated.

Problems at the Boston Housing Authority; Problems at Fidelis Way

Since the founding of the Boston Housing Authority (BHA) in 1935, public housing has been an important resource for low-income families in Boston. In the context of a very tight and very expensive rental housing market, public housing now houses approximately 10% of the city's population, one of the highest rates of any city in the United States. Like many large city housing authorities, the BHA and the housing developments it managed underwent a prolonged period of decline beginning in the 1960s, one marked by institutional corruption, physical deterioration of the housing stock, and social disruption among the residents and their neighbors. The redevelopment effort at Fidelis Way was made necessary by a combination of factors, some specific to this individual housing development and others more rooted in the broader failures of the public housing system. Although there were certainly aspects of the architectural design that contributed to the problems faced by residents, the core of the matter was a breakdown in management that exacerbated both physical and social decline. Built in 1950 under a Massachusetts program intended to provide housing for war veterans and their families, Fidelis Way generally seemed to have received high marks from its residents for the first 20 years of its existence. Lengthy interviews conducted with 41 residents in 1993 reveal that nearly all of the 17 respondents who arrived before the mid-1970s recall coming to a place that was quiet and clean with reasonably well-kept grounds (Figure 5.1). Although perusal of the Boston Housing Authority archives does reveal a number of references to problems with vandalism even in the 1950s and 1960s, for the most part these early years seem to have been comparatively trouble-free, especially given the fact that the place housed more than 1500 children.

For those who arrived in 1976 and shortly thereafter, however, the memories are much darker. The first impressions of residents who arrived in those years were dominated by images of decline: the development was described as "*dirty*" . . . "*dumpy*" . . . "*ugly*" . . . "*ragged*" . . . "*terrible*" . . . "*horrible*" with "*many empty, burned apartments.*" One resident likened the place to "*a war zone.*" Socially, several persons indicated that "*everyone kept to themselves.*" When asked to describe conditions at the end of the 1970s, just before the redevelopment effort was launched,

Figure 5.1. Commonwealth Development in the Early 1950s

residents spoke of the shame that they felt, whenever friends came to visit and had to cross the development's desecrated landscape and negotiate filthy hallways to reach their apartment. As one resident described it, although "*inside my apartment was pretty nice, outside everything had deteriorated; fences were down, the grass was gone, and the trees were broken.*" Another responded similarly, "*My apartment was OK, but outside in the hallway was like hell, with holes in the walls. There was no security in the building—the doors stayed open all the time and anyone could walk in and out.*" Those who lived on the upper floors of the mid-rise buildings invariably complained that the elevator was "*almost always broken.*" For some residents, even the inside of the apartments did not remain a haven, given the presence of mice, rats, and cockroaches.

Donald Rapp, hired in 1979 to help tenants at Fidelis Way to organize and obtain funds for redevelopment and social services, remembers his vain attempts to convene meetings of tenants during his first couple of months on the job. He couldn't understand the reluctance on the part of some of them—even key leaders—to attend these gatherings in the evenings. The problem, he finally realized, was that they lived in high-rises above totally abandoned floors, and as he puts it, "the elevators wouldn't work, and you couldn't get women to climb all they way down the dark stairs from their 5th or 6th floor apartments." To compensate, he had to carry a flashlight in his car; at night, he would hike up the stairs to the apartments, hike all the way down, all just to take them to the meeting in their own development. He remembers thinking that "this is a level of fear that I just can't imagine living with. . . . The whole development was dark and it was a very scary place."

Inside the buildings, he recalls, there were long and unexplained power outages and the constant smell of urine in the hallways and stairways. Even beyond that, "the sewerage problem was just monumental. Periodically, every few weeks, something happened and it would back up, and out of people's sink drains and toilets would pump up all the sewerage from the development and come out into the house. So I'd come in the morning and know that it happened the night before, because everybody's carpets and pieces of furniture would be hanging out on the lawn and being drained" (Rapp, 1994). This story, and others like it, confirm what a dehumanizing place this public housing project had become. Pamela Goodman, who coordinated the redevelopment effort from the Boston Housing Authority side, remarks that by 1980, the physical conditions had become "just horrendous." She notes that there was so much ice inside the buildings that "kids used to go sledding down the back stairwells. . . .

The plumbing situation created problems that no one could address, because you couldn't just keep doing bandaids. You can't imagine living in that condition" (Goodman, 1994).

The institutional causes that contributed to this decline are manifold, interrelated, and complex. In part, housing planners had failed to anticipate a future group of residents who would be a far cry from the original carefully screened homogeneous collection of small and transient veterans' families, a failure that had implications for both the physical environment of the development and for the economic viability of the public housing system (Vale, 1992; Vale, 1995). By the mid-1950s, as the programs that called themselves "public housing" changed their emphasis to accommodate the persons displaced by slum clearance and urban renewal programs, the demographics of the public housing tenant population began to change; with this, the meaning of the institution changed as well. For some, public housing had been the way station that it was intended to be. The growing strength of postwar housing programs for moderate-income groups helped many of the white working-class families to own homes of their own, sometimes even in the suburbs. Slowly, public housing became seen not as a temporary haven for an incipient middle class, but as a domicile of last resort. Moreover, it increasingly became housing for nonwhites who were lured into the cities by promises of lucrative employment that remained unfulfilled (Vale, 1992).

As urban renewal schemes cleared the way for new public housing, regulations stipulated that those displaced by slum clearance would receive first priority for rehousing in the new projects. This, in many cases, entailed a relaxation of the tenant screening processes and a reduction of average income. Increasingly, the families displaced into the projects by urban renewal efforts were large and headed by a single parent. Because rent was set as a fixed percentage of residents' income, and income was declining, project maintenance became an increasing problem, especially because of the absence of separate government operating subsidies. Yet, diminished rent collections cannot provide a full explanation for the inability of public housing authorities to manage and maintain their developments. Other factors must also be considered. As the socioeconomic status of the public housing population deteriorated and the percentage of white families housed decreased, the political clout of residents—never very great to begin with—declined further, making them (and the places they lived in) increasingly vulnerable to negligence or mistreatment. When Commonwealth was built, its expansive grounds had been planted with trees and grass, with fences to define areas for different activities and parking confined to designated central lots. As the quality of the landscape and maintenance deteriorated, many areas were paved with asphalt rather than replanted.

Given the preponderance of children, the layout of buildings led to situations where up to 50 children might live along a single corridor or share a common entry stairwell; a hundred or more might be dependent on a single elevator. Few adults were available to control them, and the children had few intact amenities to use for play areas. Crime and vandalism were an almost inevitable result, though much of this may well have been perpetrated by outsiders who viewed "the projects" as a kind of free-for-all zone.

As crime and racial tensions mounted in the 1960s, vacancy rates in many public housing developments in Boston and elsewhere rose steadily, though Fidelis Way remained at full occupancy until the mid-1970s. Beginning at that time, Boston Housing Authority (BHA) maintenance took a turn for the worse, and a cycle of vandalism, departure, and abandonment set in. In the words of one analyst, ". . . poor management by the BHA created grim conditions in many developments. Disruptive tenants were seldom evicted and repairs went undone, with employees often doing no work at all. And, because of political patronage, the housing authority became a deeply entrenched and isolated haven for the politically faithful" (Bratt, 1985). In 1975, the vacancy rate at Fidelis Way was only 2.2%; by 1978, nearly one fourth of the units were unoccupied. By late 1979, the vacancy rate had soared to 42%, finally peaking at 52% at the end of 1980. The vacancy rates and conditions at Fidelis Way during this period were not atypical of what was found in other BHA developments. By the late 1970s, the mismanagement and deterioration of the BHA had become so egregious that the Authority was placed in court receivership in 1980.

Boston Public Housing Revitalization in the 1980s: Preconditions and Catalysts

Despite sharp declines in the funds available for public housing, court-appointed receiver Lewis H. (Harry) Spence presided over something of a renaissance in public housing design and redevelopment in the early 1980s, a revitalization effort unparalleled in any city in the United States. Through creative interpretations of federal funds intended for public housing "modernization," combined with an impressive array of financing strategies involving both state and private monies, the agency assembled tens of millions of dollars with which to attempt a revitalization of a small portion of their public housing stock. Rather than divide scarce funds equally among dozens of developments, BHA officials decided to concentrate resources on some of those projects that had been the city's most dramatic failures. In part, the reason for this disproportionate allocation of funds stemmed from a realization that these projects, although still in large part structurally sound, were deteriorating to the point where they could no longer be inhabited; for the most troubled of developments, the trade-off seemed to be between saving some or losing all. The motivations were clearly political as well. As Spence put it, "If we didn't do a couple of grand, glittering showcases, the polity would have been yelling and screaming that we weren't getting anywhere, that we should just close down public housing. It was explicitly clear that we had to do these projects as a way of getting time and buying respect" (March, 1983, p. 42).

Just before Spence's arrival in early 1980, the planning department of the BHA—to its great credit—had instituted a complex and multidisciplinary evaluation system to determine where scarce redevelopment dollars should be spent. In an internal planning document titled "Site Selection Criteria for Substantial Rehabilitation," the Authority evaluated its 10 "most distressed" family developments, judged to be those having the highest vacancy rates as of August, 1979. Commonwealth, nearly 42% vacant at that time, was near the top of this list, but all 10 projects considered for large-scale rehabilitation had vacancy rates of at least 27%. There was a double reason for focusing on vacancy rates; this was seen both as a key marker of broader and deeper problems and also, paradoxically, as a helpful redevelopment precondition, because the high vacancy rate allowed for more flexibility in relocating existing residents during reconstruction.

The 10 places were then analyzed to determine "at which development might the dollar investment have the greatest impact as well as the greatest likelihood of success" (BHA, 1979b). The BHA first analyzed all 10 in terms of site accessibility, physical design, neighborhood characteristics, and tenant characteristics, because it was felt that these were important to the "potential success of any large-scale rehabilitation effort" but usually represented factors that the BHA would have "little ability to change or control." The three top ranked developments according to these initial measures were then subjected to further review of a more qualitative and development-specific nature, addressing such issues as "the political aura surrounding a site, ongoing efforts of the Tenant Task Forces, environmental impact concerns, and future neighborhood development plans" (BHA, 1979b).

Commonwealth, located in Boston's Brighton neighborhood, scored very highly in the first-stage review for many reasons: its excellent access to public transportation, its reasonable proximity to shopping and recreation, its adjacency to several hospitals, its plenitude of parking, the clear street orientation of its buildings, and its "relatively stable, well-kept residential neighborhood" (though, in fact, most of its immediate neighbors are institutions). It scored less well in terms of physical design factors, due primarily to the presumed intractability of its seven mid-rise structures, its relatively large size (fourth largest among the 10 "most distressed") and its relatively high density (45.6 units per acre), more than double the density of the least dense among the 10 projects under consideration. Socioeconomically, Commonwealth's tenants at that time were fairly typical of those in other BHA family developments, with only 14% of its adult residents employed, half of its tenants under age 18, and 94% of its families female-headed.

Based on all these initial criteria, Commonwealth ranked second among 10 BHA family developments in terms of "likelihood for successful redevelopment." When evaluated in terms of less quantifiable measures during the second level of study, Commonwealth emerged as a clear redevelopment priority for the BHA. When compared to the other two most highly ranked sites for potential redevelopment, Commonwealth was seen to have the "most urgent unmet physical needs" and "much more of a pressing community and political concern."

Although few would question the dire physical circumstances of the Fidelis Way project, the issue of "pressing community and political concern" is one that requires further examination. Ultimately, decisions about redevelopment priorities seem to hinge on such highly subjective judgments, and the "success" of the Commonwealth redevelopment effort has much to do with the events that preceded any firm decision to fund it. In addition to the BHA planning department's Site Selection evaluation, there were two other key factors. First, by late 1979 when the BHA accelerated its consideration of substantial rehabilitation options for some of its developments, Commonwealth was among the most well-documented of BHA properties, having already been the subject of two major studies.[4] The first report set out approaches for the physical and social redevelopment of the project, concluding that the development would have to be reduced to no more than 450 apartments if these apartments were to meet the state guidelines for occupancy standards, because these called for apartments that were approximately 30% larger than what existed. This report also stressed the need for management reforms, including a much greater management role for residents, and emphasized the importance of increased resident training and employment opportunities, together with improved social service program delivery.

The second report (known as "the Walsh Report") focused on financial issues, confirmed that "the Fidelis Way development, unlike other public housing projects, had excellent real estate redevelopment potential" (Walsh & Associates, 1979, p. 5). Beyond its locational assets, the report took note of the ample parking opportunities, the relatively trouble-free racial and ethnic integration of the development, the continued strength of real estate values in the area immediately adjacent to the project, and "the high degree of desire to improve the project coming from the tenants themselves" (p. 25). This report explored many different financing options for a redevelopment effort and helped to introduce into the discussion some of the mechanisms that did come to be used. First was the idea of "federalization" in which federal HUD funds could be used to reduce the mortgage on a state-owned public housing development, and the Massachusetts state money then being used for debt payments could be used instead for rehabilitation. Under this federalization arrangement (an option that now no longer exists), the development could also receive operating subsidies from both the Massachusetts and the federal governments (Seebold, 1979).

A second development strategy that seems to have gained impetus from the Walsh Report was the idea of using a turnkey approach to redevelopment. Under this process, the emptied buildings of the project would be turned over to a private developer, who would execute the rehabilitation according to a detailed agreement and then sell the revitalized development back to the BHA. By temporarily removing the development from the public sector, the turnkey process allowed the housing authority to use advantageous construction financing terms available from the Massachusetts Housing Finance Agency (MHFA), and to bypass the expense of the public bidding process and union regulations. Although the BHA had developed more than 2500 units of housing under the turnkey process between 1970 and 1981, most of this was for scattered site housing and elderly housing, and Commonwealth was to become the single largest of such efforts (BHA, 1981a, p. 1).

The Walsh Report was also extremely critical of BHA management practices, and this helped initiate discussion of private management for

Commonwealth in addition to turnkey redevelopment itself. In the broader context of a public housing project where the redevelopment by a private developer as market-rate housing was found to be "financially feasible," and where there was "a certain amount of pressure for the BHA to sell the property," much of the discussion revolved around how to obtain federal funds and private sector involvement, while preserving the housing for "low- and moderate-income" tenants (Walsh & Associates, 1979, pp. 6, 40).

In addition to the two reports about Commonwealth itself, the true catalyst for the redevelopment effort was the Monastery Hill Planning Task Force (MHPTF) that was established in June 1979 to assist the Boston Redevelopment Authority with coordinating all planning and development activities in the broader institutional neighborhood that surrounded Fidelis Way (Wallace, Floyd, Ellenzweig, Moore, Inc., 1979). Though this Task Force ultimately did not get to plan the future of Monastery Hill as a whole, it brought together representatives from all of the city and state planning, funding and regulatory agencies along with a broad array of neighborhood agencies and groups—including representatives from both the Commonwealth Tenants Association and the Fidelis-based Commonwealth Health Improvement Program—thereby providing a highly public platform for discussing the needs of the development and its residents. By giving the plight of the project and its residents high visibility in city and state government, by integrating well-organized tenants into a major coalition of powerful figures, and by incorporating the views of neighborhood residents without letting abutters gain the capacity to control the whole process, the Task Force facilitated a great deal of the groundwork for a successful redevelopment effort. By the end of 1979, the local Brighton press (which has been at all times extremely supportive of the Fidelis Way redevelopment effort) could rightly claim that, as a result of the MHPTF process, Fidelis Way was receiving "more attention" than any other housing development in the city. Because of this Task Force, BHA Planning Director William Karg acknowledged that the BHA felt "an urgency to improve the project" (Seebold, 1979). In retrospect, the BHA's own comparative study of redevelopment potential and the three planning reports—when taken together—provided a level of preimplementation planning that seems to have been of immense value to the success of the Commonwealth redevelopment venture. As a strategy for identifying where scarce resources can best be spent, and as a method for analyzing redevelopment options and mobilizing support, this process could well serve as a model for housing redevelopment efforts elsewhere.

With so much background work already completed, the arrival of Harry Spence and the commencement of the BHA receivership in the spring of 1980 brought with it the possibility for a funding mechanism. According to the analysis of a local reporter, "Because of the joint efforts of the community and tenants, which have never been so closely allied before, the receiver was hard put not to recognize Fidelis Way and the opportunity there" (Seebold, 1980b). Spence, who served as the BHA's Receiver and Administrator from 1980 to 1984, recognized the potential and directed a tremendous amount of energy and discretionary budget allocations toward the realization of this redevelopment effort. Ultimately, $20 million of the funding for the redevelopment came from unexpended HUD development funds already committed to the BHA, augmented by $6.2 million more from HUD and $5.3 million from the state of Massachusetts (NCSDPH, 1992, pp. 4-8, 4-36).

■ Redevelopment: Revitalizing the Tenants, the Buildings, and the Management

The $31.5 million transformation of the Fidelis Way housing project into Commonwealth Development is a story of a strong core of residents, activated with the assistance of a hired "community organizer," and backed by a strong network of neighborhood, city, and state organizations, who jointly worked with the local housing authority and a private development company to negotiate the terms of the physical and social goals for the redevelopment, as well as the means by which the redevelopment effort could be managed and sustained. At base, it is a story about the process of building a mutual level of trust and respect among parties that had previously been perpetual adversaries. Residents, working with skilled architects, landscape architects, and urban designers—as well as with the housing authority redevelopment team—agreed on the priorities for the physical redevelopment and helped negotiate a 223-page Management Plan that established a binding agreement covering a wide variety of community rules and procedures and the agreed upon methods for addressing any violations of these norms. The design goals outlined in the Developer's Kit, drawing on the CPR study and considerable input from BHA staff, called for a reduction in the number of apartments, an increase in their spatial standards, and a variety of measures to make the buildings more responsive to the specific needs of all segments of the public housing population. In short, the redevelopment effort involved a drastic overhaul of the physical character of the site, combined with a wide-ranging and serious attempt to introduce and enforce a management and maintenance plan that could sustain the development as a secure and attractive community.

Revitalizing the Tenants

Throughout the process, the residents of Fidelis Way remained at the core of the attempt to revitalize the physical environment and management practices. Engagement by the residents and with the residents preceded both design and management reforms and ultimately helped to shape them in significant ways. Redevelopment was not something that simply happened to these people but was something that was actively fought for by many of them. Although it may seem, in retrospect, almost inevitable that Harry Spence would have been drawn to Commonwealth as a priority, this seems so only because so many different forces converged to support it. Without active advocacy by tenants clamoring for redevelopment, attention and money could well have gone elsewhere. From the real estate development discussions of the Monastery Hill Planning Task Force, Fidelis Way residents learned that their place of residence, too, had great value in the eyes of others. Although this may have led to an initially persistent fear that the development could be sold out from under them, it also served as a confidence-building measure. Tenants who once in early meetings of the MHPTF sat silent, while their very existence was berated by certain neighborhood residents who saw them as little more than another obstacle to more "open space," became active participants in a process where fancy-titled and politically connected persons came to regard their needs as central to multimillion-dollar deals. In 1980, when the newly appointed Receiver made his first official visit to Fidelis Way to meet with tenants and view a video documenting the horrendous conditions in the development, he came away "incredibly excited." "People are getting angry again," Spence commented. "Therefore, there is more hope for Fidelis Way" (Seebold, 1980a).

By the time of the receivership, this tenant anger and frustration already had at least two productive channels for action, the Commonwealth Health Improvement Project (CHIP) and the Commonwealth Tenants Association (CTA). The creation of CHIP in 1978 helped facilitate tenant organization even before the rest of Boston converged on Monastery Hill. Sponsored by the Allston-Brighton Medical Care Coalition and initially funded through the United Way, CHIP began as an on-site health referral agency, aimed at preventive care (Allston-Brighton Medical Care Coalition, 1978). Yet, CHIP's initial encounters with the tenants made it clear that the "residents did not consider health, in the strict sense of the word, one of their priorities." To the tenants, CHIP's staff soon realized, "health started with security and safety, personal and community sanitation, adequate housing, financial security, employment, education, recreation, and access to quality services from public and private institutions." Because of this, CHIP rapidly evolved into a much more wide-ranging organization, with information, referral, and advisory activities that attempted not only to solve tenants' health care problems but also to attend to their housing, security, employment, education, day care, legal, and recreational needs (CTA and CHIP, 1980, pp. 2-3). As the Fidelis Way redevelopment effort gained momentum, CHIP added a second "H" to its acronym and became known as the Commonwealth Health and Housing Improvement Program (CHHIP).

When combined with the Commonwealth Tenants Association, Commonwealth's tenants enjoyed a kind of double form of representation, both on the Monastery Hill Planning Task Force and in the eyes of the BHA. The CTA was, by the time the redevelopment began, already one of the more active of such organizations in the city, and was also one of the few with a multiracial and multiethnic active core membership. Established in 1969 as part of a citywide initiative in public housing projects, it became incorporated and was granted tax-exempt status in November, 1979; this enabled it to seek foundation grants more actively and directly. In 1980, the CTA began publishing a bilingual monthly newsletter, *Fidelis Way Speaks . . . Fidelis Way Hablas . . .* that not only provided updates on redevelopment progress but served as a forum for concerned residents to discuss openly the kinds of high standards for resident behavior that ought to accompany the architectural and managerial changes. The CTA, then as now under the leadership of resident Bart McDonough, brought together a core of tenants whose central activity for years has been ensuring the success of the redevelopment. As a result, resident voices could be heard at all key decision points of the redevelopment process, whether regarding design or management. More important, at Commonwealth, this input became institutionalized to the point where it was both anticipated and legally guaranteed.

Revitalizing the Apartments, the Buildings, and the Site

The BHA's careful, thorough, and ambitious "Developer's Kit" for Commonwealth, issued in June 1981, was of central importance for defining the parameters of the resultant redevelopment effort. It required at least 390 units of housing divided into a carefully articulated unit mix of elderly housing and family housing in apartments of various sizes based in large part on the bedroom requirements of the existing population. It proposed consolidation of housing for the elderly into two adjacent mid-rise buildings with a community building connecting them. It called for the transformation of low-rise buildings into duplexed or triplexed townhouse-like apartments with individual entrances and private yards for large families, and the reconfiguration of the lower two floors of other mid-rise buildings into similar duplexes for large families. It suggested the reduction of density through selective demolition of a mid-rise building and its replacement with a community center and management building, and it insisted on the articulation of "defensible space" in public areas. In addition to these and other basic design and programming matters, the Developer's Kit contained a wide variety of lengthy appendices, including a very detailed set of "Planning and Design Criteria." These included recommendations for grade changes in the landscape to permit all private entrances to be entered from ground level (to make for "a residential feel and clear delineation between semi-private and public space"), noting that these newly configured dwellings must feature "clearly defined private space at the unit edge of not less than 200 to 300 square feet." The Kit called for parking to be located in small lots where each car could be viewed from its owner's apartment and insisted on a "clear territorial definition of open space areas . . . attractively and sensitively designed with play areas, sitting and picnic areas, and planting." In short, the Kit mandated that "every portion of the site must be dealt with and programmed," while noting that the "design must reflect [the fact that] the primary users of the site are children."

Again and again, the clear message given to potential designers and developers was that "care must be taken in design and material selection to avoid an institutional and repetitive design." To accomplish this, the Developer's Kit proposed that "each building and unit should have a definite front and back. Fronts are typically associated with the street, where cars are parked and where units are entered. Back doors should also be provided for individual units not only for convenience but also to help establish private territory at the back of the buildings. The goal is to define as much of the building edges as possible as private territory associated with a particular unit in order to discourage loitering in these areas. Buildings should be paired so fronts face fronts and backs face backs." The effort should be, in short, "to convert the low-rise walk-up buildings to row house-type structures" (BHA, 1981a).

Six teams submitted bids, and the winning proposal by developer and contractor John M. Corcoran and Company and architect Tise Associates was judged to have excelled in a number of the most important design criteria articulated in the Kit. According to the selection panel, in addition to its competitive price and the fine reputation of the team members, the design was favored for many reasons:

> All common areas are eliminated in the low-rise buildings that will contain most of the large families. Individual access also is provided for a number of large family units on the first two floors of the mid-rises and results in the lowest number of elevator-accessed family units proposed. These aspects will substantially bolster security and the livability of both the individually accessed apartments and the elevator apartments. The relationship of buildings to site is also particularly strong. Pedestrian pathways are clear and safe, parking spaces are both convenient to and visible from units, and open space is well programmed for maximum tenant identification and responsibility. Security and maintenance problems should be reduced as a result. The proposed design is not fancy or elaborate but rather a straightforward application of housing development design fundamentals (BHA, 1981b).

As one part of the thoroughly commendable selection process, involving wide community consultation and careful scrutiny of proposals by the BHA and various Massachusetts state agencies, the Commonwealth Tenants Association hired a consulting architect, David Lee, who spent thirty hours ensuring that residents would be "fully informed as to the content, significance, and implication of the technical aspects of the six proposals" (CTA, 1981).[5] As part of what tenants called a "painstaking review" of each proposal's drawings, the architect constructed scale-model pieces of furniture to help them envisage life in the redeveloped apartments. The results of this independent review by the Commonwealth Tenant Association board and other interested residents concentrated on design and programming issues. For residents, the Corcoran proposal was "the overwhelming favorite"; they gave it three grades higher than the next highest-ranked submission.

Though the winning scheme followed closely the demands of the Developer's Kit, the Corcoran/Tise team did make significant advances over the Kit, both in terms of programming and design. Before redevelopment, two thirds of the apartments were in six-story buildings, with 30 families sharing an entrance and single elevator. Under the redevelopment plan, two of these seven buildings were demolished to reduce the actual and perceived density of the site, to be replaced by a low-rise daycare facility and a management and community building. The single street looping through the development was reconceived in two parts, one a more direct route through the development that served as a more public way and ran past the new community buildings, the other a more private loop intended for the use of residents and their visitors (Figures 5.2, 5.3, 5.4, 5.5, and 5.6).

The perceived mismatch between families and high-rises was addressed in a couple of different ways. As proposed by the Developer's Kit, two of the six-story high-rises were reconfigured to accommodate elderly residents in small apartments and connected at the ground floor by a new

Figure 5.2. Commonwealth After Redevelopment: A view taken from the same vantage point as Figure 5.1, but 44 years later, shows the ways that redevelopment has reduced the density of buildings, reconfigured the road and path system through the site, and reconnected the buildings to the landscape.

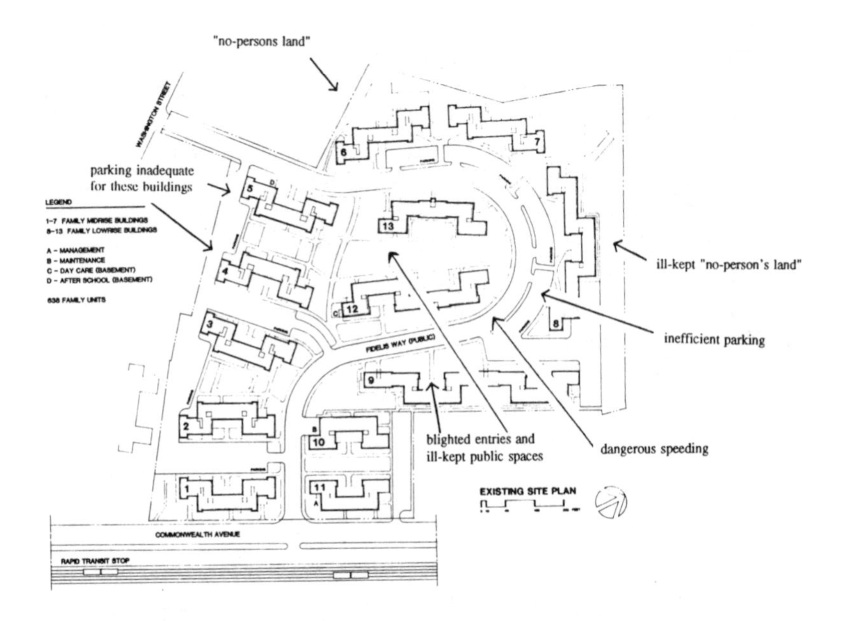

Figure 5.3. Site Plan Showing Problems Before Redevelopment

structure housing community facilities for the households in those buildings. Following a suggestion in the Developer's Kit, three other high-rises were reconfigured so as to minimize the number of children living in elevator buildings by redesigning the buildings so that the bottom two floors housed the largest families and functioned as two-story townhouses, reached by private exterior entrances to each apartment. As part of this retrofitting of rowhouses into the base of high-rises, the number of families relying on elevators was reduced and the number of families able to have private back gardens was correspondingly increased. The three-story walk-up buildings that made up the rest of the site were similarly reconfigured into two-story and three-story rowhouses with private entrances and gardens. These new zones of privacy and control were greatly reinforced by careful attention to landscape.

At Commonwealth, principles of defensible space adapted from the work of Oscar Newman (1972) were used in particularly innovative ways,

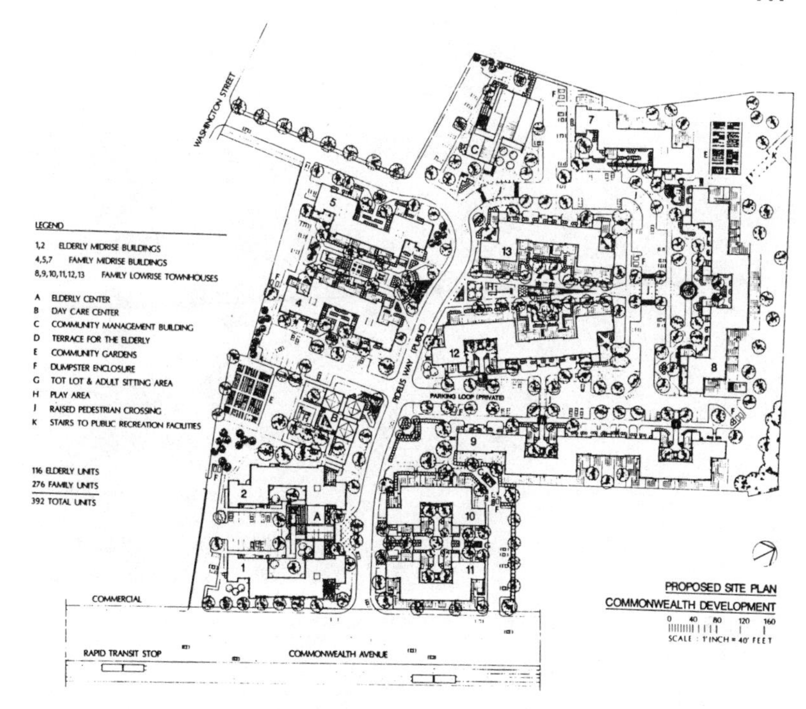

Figure 5.4. Site Plan After Redevelopment

involving elaborate regrading of the site to define zones of private and semi-private space within the development. This aspect of the redevelopment is one part of a broader effort to support the BHA's effort to make public housing look less alien, to reconcile it with the preferred American domestic living type: the single family home with its own private yard.

Although management of traffic and parking at Commonwealth contributes to the definition of hierarchies of public and private space on a large scale, the landscape treatment fosters these in smaller and more subtle ways. Plantings and gradation changes foster degrees of privacy by creating a variety of traversable, semitraversable, and nontraversable barriers. By subdividing all "common areas" into spaces with a clear function and an identifiable set of users, Newman argued that families in public housing would choose to "defend" all parts of the housing development, thereby reducing crime, vandalism, and the need for maintenance. Newman's ideas, consistent with much of what has been described

Figure 5.5. Townhouses at the Base of Six-Story Buildings After Redevelopment

Figure 5.6. Relandscaping to Create Zones of Private and Semi-Private Space

here, were aimed at encouraging project residents to regard their homes as extending into a variety of semi-private areas surrounding them. Although this is hardly accomplished by landscaping alone, the replanting and regrading have certainly worked toward this end.

At Commonwealth, what was once a series of apartment blocks with common entries eight steps (5 feet) above grade was transformed into 176 duplexes and triplexes, each with a private entrance at grade, insulated from the sidewalk by a gentle slope of grass-covered fill. This thoughtful regrading created a clear zone of semi-private space that visually belonged to the townhouse behind it. At the rear, in many cases, the change of grade also allowed the private backyard or patio to be on a different level from the semi-private and semi-public play areas that were defined beyond it. Consistent use of curbing in other, more public, areas seems to have held up as a barrier that most people will not violate, even in pursuit of a shortcut.

Just as residences are given a kind of semi-private buffer zone, so too play areas for children of different ages are carefully delineated. There is an emphasis on dispersion, diversification, and safety, and the facilities are juxtaposed with residences and with each other to meet the requirements for surveillance by both parents and older children. Moreover, implicit in the design of play facilities in all of these developments is a recognition that children will inevitably wish to use areas for play that are not purpose-built by architects. Rather than fight this tendency, the architects sought to accommodate it, with extra-wide sidewalks and the provision of a variety of both hard and soft surfaces. From parking lots to tot lots, the planners of Commonwealth followed the BHA directive to regard every square foot as having a potential use and a potential user. Beyond such technical matters, however, attention to landscape seems to have played an even more important role in rebuilding trust between residents and management. One of Corcoran's first acts upon taking over Commonwealth was to plant flowers in the public areas of the development. More than a decade later, at a place where flowers continue to be planted and carefully maintained, this first simple act seems to be lodged in the memories of residents and housing authority officials as a symbol of a new kind of respect.

Taken overall, the approach to redesign and tenant involvement undertaken at Commonwealth would seem to affirm the value of bringing all parties together to negotiate a clear and comprehensive document that establishes principles and sets redevelopment goals right from the start. Moreover, in addition to the necessity of a high-quality Developer's

Kit, what seems most salient in the Commonwealth process is the extent to which tenant decision making was able to remain continual, well-informed, and legitimately influential. Given that tenants would be living in and with the resultant housing, this central role seems essential.

Revitalizing the Management

In addition to these design ideas, many of which were also simultaneously or subsequently implemented in one form or another in other public housing redevelopment efforts in Boston and elsewhere, one aspect that was unique at Commonwealth was the decision to privatize the management. It had already been an unusual move for the BHA to contract with a private developer through a turnkey process to carry out the actual reconstruction, but this further accommodation with the private sector was unprecedented for a large family public housing project. Though it was the BHA's decision to go with private management, it was a decision that tenants wholeheartedly endorsed. According to Don Rapp (1994), the CTA Executive Director at the time, the BHA was "looking for tenant support, and got it. The tenants also had no confidence the BHA would manage the place well. It was not controversial." Beyond the shift to private management, the tenants also wanted to play a substantial role in management themselves—though they evinced little interest in taking on responsibility for day-to-day operations (CTA Management Committee, 1981).[6] MHPTF co-chair Trietsch (1994) confirms that this commitment to enhancing the clout of the tenants was widely shared: "We knew that if it wasn't tenant-controlled management or unless the tenants had a significant role in the management of the project, the chances are it wasn't going to succeed. We weren't going to get the tenants to buy in to the work they were going to need to do to see the thing through."

At Commonwealth, the CTA entered into a legally binding agreement for a management plan with the Boston Housing Authority and a private firm, the Corcoran Management Company (CMC), a division of the firm that had served as the developer, though CMC was chosen to take on the management aspect only after a separate competitive bidding process in 1982 (BHA, 1982). The agreement sets unusually high standards for management and maintenance and even allows the tenants to fire the management with a 30 days' notice (CTA, BHA, & CMC, 1983). The Management Plan addresses a broad range of issues and, more than a decade later, remains a reference point for considerations of tenant selection, rent and utility allowances, charges associated with damage to apartments and repairs, community rules, eviction policy, maintenance (including a detailed account of CMC's preventive maintenance program, corrective maintenance procedures, and standards for janitorial and grounds maintenance), resident services, tenant orientation, security, personnel policy regarding the hiring of residents, the budget, and the lease. The interviews with residents confirm a very high degree of satisfaction with the implementation of this plan. For example, fully 90% report having no complaints about the responsiveness of maintenance personnel, a figure far higher than that found at four other Boston public housing developments that were surveyed (Vale, 1994).

Even more important, the genius of the Commonwealth Management Plan is perhaps its ability to instill a sense of coresponsibility for the quality of upkeep and behavior in the development. Because residents had themselves helped to formulate the rules and Corcoran Management Company is willing and able to enforce them, any departure from the high standards that both management and residents have set can be met with widely supported sanctions.

Measuring, Explaining, Sustaining, and Replicating "Success"

Measuring "Success"

When one asks a wide range of individuals associated with the redevelopment to evaluate its success, there is an almost unanimous impression that the transformation has succeeded in many different ways. Though some tenants, particularly teens, voice some concern that the stricter enforcement of community rules has been too draconian, nearly everyone else praises what has happened here.

Surveys of residents carried out in 1979—near the nadir of the predevelopment conditions and again in 1993 after nearly a decade of reoccupancy—suggest something of the magnitude of change that has occurred.[7] In 1979, when a diverse sample of 80 adults were asked whether Fidelis Way was "a good place to raise kids," 78% answered that it was not. Fourteen years later, 78% of a similarly diverse sample of adults in more than 10% of the households living in the redeveloped Commonwealth responded that the development *was* "a good place to raise kids" (Table 5.1).

TABLE 5.1 Is this development a good place to raise kids?

	1979	*1993*
Yes	19%	78%
No	78%	17%
Don't know or no answer	3%	5%

Similarly, whereas in 1979 when respondents were asked, "Are there good places for children of all ages to play?" only 19% believed that there were, by 1993, this figure had ballooned to fully 90%, a clear endorsement of the landscaping efforts, the private yards, and the rehabilitation of Overlook Park that adjoins the development (Table 5.2).

TABLE 5.2 Are there good places for children of all ages to play?

	1979	*1993*
Yes	19%	90%
No	75%	10%
Don't know or no answer	6%	—

In terms of management and maintenance issues, the contrast between preredevelopment and postredevelopment is equally stark. In 1979, three quarters of the respondents complained about the inability of maintenance staff to address problems in their apartments; in 1993, only 10% indicated that they had experienced any problems with maintenance (Table 5.3).

TABLE 5.3 Have you had any problems with maintenance?

	1979	*1993*
Yes	75%	10%
No	25%	90%

Because this measure was taken nearly a full decade after the completion of the redevelopment and the survey was conducted during a time when the Corcoran Management Company argued that the Boston Housing Authority was providing Commonwealth with a disproportionately low subsidy, the relatively low level of complaints, combined with a superb record in the turnaround of work orders, surely provides compelling testimony that Commonwealth's management is now in very good hands.

One final measure of the redevelopment's success is a clear indication that residents now feel far safer in the development than they did in 1979. Even with the dramatic acceleration in drug activity and drug-related violence that has devastated so many low-income communities since the mid-1980s, the survey results from 1993 show a marked increase in security since the redevelopment. In 1979, when respondents were asked "How safe do you feel" at Fidelis Way, only 11% indicated that they felt "very safe." Nearly two thirds said that they felt "somewhat unsafe" or "very unsafe." In complete contrast, the 1993 survey results—incorporating both daytime and night-time assessments—show that fully 73% of respondents say that they feel "very safe," an additional 22% report feeling "somewhat safe," and only 5% say they feel "somewhat unsafe" or "very unsafe" (Table 5.4).

Despite such encouraging findings, however, it must be said that Commonwealth is hardly immune from the problems confronting inner-city America. Of the Commonwealth respondents in 1993, 54% indicated that they believed drugs to be "a major problem" at the development, and more than 80% said that they believed the problem to be "getting worse" or "staying the same" since 1990. Nevertheless, most contended that the drug problem was worse in surrounding neighborhoods than it was in the development itself. This finding would seem to challenge the views of those who contend that much of the success of the Commonwealth redevelopment was due to its location in a relatively stable, economically

TABLE 5.4 How safe do you feel in this development?

	1979	*1993*[8]
Very Safe	11%	73%
Somewhat Safe	27%	22%
Somewhat Unsafe	41%	4%
Very Unsafe	20%	1%

diverse, and racially and ethnically mixed community. Although it is certainly true that the Allston-Brighton neighborhood is indeed relatively less troubled than those surrounding many other public housing developments, is better serviced by public transit, and contains many more desirable amenities, such as health services, this area is not without its troubles. The redevelopment effort, for instance, was opposed by some neighborhood residents who would have preferred to have had Fidelis Way demolished and its residents dispersed. Yet, in contrast to many public housing developments in Boston, where socioeconomic conditions in the surrounding neighborhoods are as depressed as the conditions within the public housing project, it is a very different redevelopment task when the housing project is seen as the neighborhood's chief problem rather than the other way around.

Explaining "Success"

In the end, there seems to be no consensus about which aspect of the redevelopment was the most essential, or whose role was the most central. Planners stress that the Monastery Hill planning process was the crucial catalyst, emphasizing the need to build a coalition of powerful players before large-scale redevelopment dollars for a low-income community could be obtained. Designers seem to believe in the paramount value of their own contributions. They point to the reduction of density on the site, the reconfiguration of buildings and outdoor spaces to reduce the number of hallways, the increase in apartment sizes and the number of private entrances and private yards, the redistribution of elderly residents and young families into units and building types better suited to their needs, the regrading of the site, the creation of multiple age-specific play areas, and the programming of new community facilities.

Many of those involved in the management of Commonwealth, however, suggest that the place could have been just as successful with a lower investment in design costs, particularly such amenities as the elaborate community facility. They recognize ways that management tasks have been eased by design changes (such as the reduction in the number of unsupervised hallways and young children living in elevator buildings with no place for play), but note that the overriding issue is the quality of the management (Corcoran, 1994). Some claim that management quality is closely tied to the ability of the private sector to circumvent the rules that bog down government bureaucracies. By hiring on-site maintenance personnel who are skilled at many trades, without having to work through a central housing authority that is working under union regulations, they argue that a responsive maintenance system is much easier to establish and maintain. When the removal of political minefields, such as union hiring, is combined with the added capacity to have service personnel closely tied to the Commonwealth community (either because they are themselves residents, or because they have come to know the residents well through sustained involvement), the relative advantage of private management is further enhanced. A sometimes heated debate remains over how much of the successful management experiment at Commonwealth was driven by its privatization. Although there seems to be little or no criticism of the performance or cost effectiveness of Corcoran Management Company, many are reluctant to see their qualities as too closely bound up with the fact that they are a for-profit development company rather than a public agency. Good management, these critics argue, means the same thing, whether it is done by government or the private sector.

In assessing the success of the Commonwealth redevelopment, all parties concur that the Commonwealth Tenants' Association, together with the community organizer who assisted them, were also central. What is most important, perhaps, is that the tenants who lived through the redevelopment effort themselves believe this. When asked to name who played the most important role in the redevelopment efforts, the vast majority credited "tenant leaders" or "all residents." A few mentioned architects, one person credited everyone equally, but nobody singled out either the Boston Housing Authority or the private developer as the key player.

Sustaining and Replicating "Success"

By many measures and according to many different explanations, the redevelopment effort at Commonwealth has been a success. Sustaining it and replicating it elsewhere poses further challenges. Though the exterior of the development has held up well under careful maintenance for a full decade, the staff of the Corcoran Management Company insist that the long-term maintenance needs of the apartments are being underfunded by the Boston Housing Authority; CMC, they contend, is being asked to manage Commonwealth with *half* of the BHA's average per unit allocation of funds (Pickette, 1994). BHA Deputy Administrator Bill McGonagle counters this by asking, "Should a development that's had a $38 million facelift receive as much of the operating budget as a development of similar size and age that hasn't received the same amount of capital improvements?" He notes that efforts are under way to reconsider the Commonwealth monthly allowance (Kahn, 1993). Although there is, of course, considerable validity to the BHA position, the danger remains that with chronic underfunding, the conditions at Commonwealth will be allowed to decline. Though rent collection remains near 100%, and Corcoran currently generates about $20,000 per month in excess rents over expenses (over and above their management fee), this money is returned to the BHA rather than reinvested at Commonwealth (Pickette, 1994).

A second concern over the sustainability of the Commonwealth transformation centers on the process of tenant selection and screening. During redevelopment, as a result of an arduous process of negotiation with the BHA, the tenants and their legal adviser (Leslie Newman) managed to obtain a comprehensive rehousing agreement that maximized the chances that individual tenants would be able to move into the redeveloped units, even if they had substantial rent arrearages. As another part of the rehousing plan, tenants were offered a one-time payment and a Section 8 voucher usable to subsidize their rent elsewhere. Most of the well-known troublemakers chose to take advantage of this opportunity (Rapp, 1994). Some senior citizens also took advantage of a one-time opportunity to move into a housing development for the elderly (NCSDPH, 1992, pp. 4-24). Those who wished to remain and be a part of the "new" Commonwealth had to agree to the terms of a new orientation (that not only explained all the new appliances in the apartments but emphasized that rules of conduct would be enforced, including the prohibition against pets). In addition, all who wished to return had to have their furniture fumigated (to reduce the possibility of further insect infestations), and—most onerously for many—had to commit to a repayment plan for rent arrears that in a great many cases was quite considerable (Pickette, 1994). Taken overall, 58% of the families residing at Commonwealth in 1981 returned to the redeveloped site (NCSDPH, 1992, pp. 4-24). The net result of all this was a postredevelopment population that was cleansed of many of its most problematic families without resorting to a high number of evictions and without systematically restructuring the occupancy in favor of families of higher income as has been proposed so often in public housing elsewhere.

In other ways, the preredevelopment and postredevelopment populations remained quite similar. Since the mid-1970s, Commonwealth has housed a majority of nonwhite families; as of 1993, Commonwealth's residents were approximately 40% white, 38% black, 15% Hispanic, and 7% Asian (BHA, 1993). And, as at most public housing elsewhere,

Commonwealth's households have been for the last 20 years overwhelmingly female-headed, single-parent families. The redeveloped buildings dedicated to elderly occupancy have certainly skewed the average age of occupancy, although even during in the early 1970s, as many as half of Commonwealth's households were headed by persons over age 60 (BHA, 1979a).[9] Perhaps the biggest demographic change over four decades has been the decline in the number of children living at the development. Whereas BHA records from the 1950s make reference to 1500 children, there are currently only about 400 residents under 18, due both to changing family patterns and to the loss of units associated with the redevelopment. Although it seems likely that the long-term decline in the density of children has made management of the development easier, since redevelopment, the focus of concern by residents and management alike has been on efforts to insure that new arrivals are carefully screened.

Between 1984 and 1988, after the development reopened, Commonwealth had its own waiting list and drew new tenants from this rather than a citywide list. As part of this, Corcoran Management Company and the Commonwealth Tenants' Association could exercise considerable control over who was admitted, including extensive background checks and home visits. Since 1988, when the BHA entered into a Voluntary Compliance Agreement with the U.S. Department of Housing and Urban Development (HUD) that prohibited such separate lists on the grounds that they could be used to foster racial segregation, Commonwealth (despite its being perhaps the most racially and ethnically diverse development in the city) has had to accept tenants off the citywide waiting list and has had no more than an advisory role in keeping out those deemed undesirable. One former manager of Commonwealth puts it this way:

> They say in the [BHA] Occupancy Department that everybody wants to live at Commonwealth. What we end up doing is getting a lot of the "trash," if you will, from across the city. It's really unfair to long-term residents there who came in under a different standard. . . . Right now, at least on a temporary basis, Commonwealth is not getting any choices about who they can house. That's what it all spreads from. It's a matter of acceptability. People at the site are going to accept a certain level of performance from us, from their neighbors, and from everyone else, and once that starts to decline, it gets to be a problem. (Young, 1994)

Moreover, in part because of its reputation as a safe and well-managed place, Commonwealth has received a disproportionate share of internal citywide emergency transfers, persons whose lives were thought to be in danger in other public housing. Often, say CMC staff, new problems accompany such people:

> When you have these internal transfers from across the city and someone would be out front on the steps drinking beer or throwing their trash out the window from the sixth floor, the people at Commonwealth are all over [Site Manager] Jim Reed like a cheap suit. They want him to get these people out *now.* (Young, 1994)

Having worked hard to achieve the redevelopment, Commonwealth residents have come to expect a higher standard of behavior. As of the mid-1990s, the need for evictions has been relatively small, averaging only a handful per year (Reed, 1995).

To date, however, this higher standard of living in terms of behavioral expectations has not been translated into economic terms. BHA figures show that in 1979, 19% of families then living in Fidelis Way reported that their major source of income came from employment; 15 years later, this figure remains at about 20% (BHA, 1979a; NCSDPH, 1992; BHA, 1993). Although this is actually among the *highest* employment rates at a BHA development and would be even higher if the 132 households headed by persons over age 62 were not factored in, it is a reminder that public housing at Commonwealth still serves a population that is extremely economically disadvantaged. As of 1993, the majority of households at Commonwealth—whether elderly singles or large families—reported incomes of under $10,000. As of 1993, with rents pegged at 30% of income, approximately 60% of households paid less than $150 per month in rent for their apartments, and fully 85% paid less than $350 per month (BHA, 1993). Clearly, given that Boston remains a highly expensive market for rental housing, Commonwealth Development stands as an extremely important source of affordable housing. Even for families living at the development who report that their primary earnings come from employment, there seem to be few market-rate housing opportunities that could provide anything like the spaciousness and quality management that is possible at postredevelopment Commonwealth. Turnover at Commonwealth is now quite minimal, usually involving no more than a half dozen apartments per year (Reed, 1995). Commonwealth Development has always, for more than 40 years, provided "affordable housing"; since its redevelopment, it has resumed providing housing that is also both decent and desirable.

Life at Commonwealth, however, remains far from paradisiacal. The nature of this redevelopment approach—centered on design and management reforms—has fallen far short of meeting the full range of tenant needs. State funds supported a wide array of programs through the Commonwealth Tenants Association (CTA) during the late 1980s—including a family daycare program and an employment and training outreach worker—but budget cuts forced staff layoffs and program cancellations after 1989 (Braverman, 1989; NCSDPH, 1992, Chapter 4). Despite the intermittently successful fund-raising efforts of the CTA and others, any kind of substantial and sustained investment in education and job training is still absent. As one newspaper article put it in late 1993, "Commonwealth's social-service safety net is hanging on by a thread" (Kahn, 1993).

If, in the end, one measure of success in American public housing is the rate at which its residents become economically able to leave it, Commonwealth's success is more limited. If, on the other hand, success in public housing redevelopment is measured by the enhancement of an attractive, safe, and stable community, where even many of those who can afford to leave will choose to stay, then the effort at Commonwealth warrants the highest praise. Moreover, because Commonwealth's success in these terms was achieved through tenant activism, responsive design and effective management—*without* resorting to the income mixing so often touted as necessary to revitalize public housing—it raises important questions about current policy trends.

Despite the ongoing problems with inadequate operating funds, tenant selection, drug activity, and the lack of economic incentives and opportunities for residents, no other redevelopment effort in Boston (and few, if any, elsewhere in the United States) has matched or sustained the range of successes found at Commonwealth. Replicating even this incomplete agenda of initiatives has proved to be difficult. One major barrier is a financial one: This sort of redevelopment is extremely expensive—more than $80,000 per unit in 1983 dollars (NCSDPH, 1992, pp. 4-36). To replicate it in a high-priced construction market like Boston's could entail well over $100,000 per apartment in total redevelopment costs. Although HUD's new Urban Revitalization Demonstration program does provide funds on this scale, the availability is extremely limited (Vale, 1993).

Others criticize the Commonwealth redevelopment effort, because it entailed a net loss of about 40% of the original units. In the context of long waiting lists for public housing and an overall shortage of affordable housing, strategies that reduce density without providing low-income replacement housing may be seen as ultimately shortsighted, because replacement housing remains politically very difficult to find or build.

Beyond these issues, the nature of Commonwealth's management experiment creates tensions within Public Housing Authorities over the usurpation of their management role by a private firm. In Boston, the BHA has so far been unwilling to let Corcoran Management Company try to manage another redevelopment effort. Indeed, spurned by the BHA and other local housing authorities, CMC has recently entered into an agreement to advise the government of Mongolia on management issues in public housing redevelopment there (Pickette, 1994).

In the end, the redevelopment effort at Commonwealth would seem to offer a number of lessons for attempts to preserve public housing elsewhere as a viable and desirable source of affordable housing. First, notwithstanding the fact that the Commonwealth turnaround was an expensive, grueling, and lengthy process, it must be acknowledged that this particular housing development had many *advantages* not present in the majority of severely distressed public housing communities. It was precisely because Commonwealth was seen as "most likely to succeed" that it was selected for redevelopment in the first place. By contrast, another BHA comprehensive redevelopment effort of the 1980s—that received similar per-unit dollar investment but was seen by the BHA as a much more risky investment—has fared much less well (Vale, 1994). The success of Commonwealth affirms the value of extensive preredevelopment planning studies to determine where scarce dollars can best be spent. Given the vast scale of investment that seems needed to effect meaningful and sustainable change in severely distressed public housing, Commonwealth's success suggests that it may make most sense to focus attempts to salvage public housing on those developments that, although they are among the most distressed, are seen as the best off among that group.[10]

However, before giving up on those places that seem to be at the margins of salvageability, whether in Boston or elsewhere, it would seem worth trying more of the techniques that worked so well at Commonwealth. The defensible space approach to building and site redesign seems to have contributed substantially to resident security, and the process of resident input into the redesign through the use of their own architect-advocate seems wholly commendable. So, too, the coalition building that occurred in the course of the Monastery Hill neighborhood planning study usefully enhanced the cause of affordable housing. Because it gave low-income public housing residents a place at the table with powerful actors from around the city, it provided an extremely public forum for reaching consensus over conflicting redevelopment priorities. And certainly, the success of the private management experiment at Commonwealth may have some broad applications, though one must be careful not to interpret this story as a blanket recommendation to privatize. It is not that a public agency merely devolved its responsibilities onto the private sector; it is that a highly intelligent group of thoughtful and motivated housing authority officials, working with a core group of committed and well-advised tenants, jointly developed a vision for a tenant-monitored system of private development and private management. Commonwealth's success entailed a carefully considered and scrupulously monitored process of finding not only an available private sector alternative but the best possible one among many to meet a variety of goals and to work in partnership with a variety of constituencies. It was not the act of privatization itself, but the hundreds of hours that went into reaching consensus on the thousands of details that went into both the Developer's Kit and the Management Plan that laid the groundwork for Commonwealth's sustained successes.

Even in the absence of more involvement in American public housing by high-quality private management companies, it may still be hoped that more public housing authorities will be able to learn from some of their methods and match their ability to treat public housing tenants like everyone else. This, at base, means developing a consistent set of expectations for behavior in return for a consistent and reliable provision of services. As John Corcoran puts it, "It's mutual respect. We listen to the tenants—we have to, because they can fire us if they're not happy with us. At the same time, we plant those flowers, and not one pansy gets stepped on" (Hanafin, 1989). In the American context, where public housing serves only a small minority of citizens and where public housing residents and advocates must constantly struggle to overcome the stigma associated with life in the projects, revitalizing and normalizing public housing communities will not be easy. Although success in the many dimensions described here is an arduous and expensive process, the story of Boston's Commonwealth Development shows us that it is possible to achieve.

NOTES

1. I wish to express my appreciation to Bart McDonough and the Commonwealth Tenants Association for helping to facilitate the interviewing process at the development,

and to Donald Rapp for loaning out his extensive files on the Commonwealth redevelopment process.

2. The name "Fidelis Way" comes from the street that loops through the project, whereas "Commonwealth" is the name of the major roadway that passes by the south side of the development. Commonwealth Avenue is a widely known east-west artery that extends from one of Boston's most upscale neighborhoods (the Back Bay) out to affluent suburban areas; as such, it clearly had a kind of cachet more in keeping with the goals of the redevelopment effort. Commonwealth had been the development's "official" name since the 1950s, and the tenants' desire to return to this name—and thereby to dissociate their new homes from the decline that had befallen Fidelis Way—seems to have been widely shared (Rapp, 1994).

3. In 1992, Commonwealth was featured as one of only four public housing turnaround efforts nationwide deemed worthy of inclusion in a book of case studies sponsored by the National Commission on Severely Distressed Public Housing (NCSDPH, 1992). The redevelopment effort has also received an Urban Design Award from the Boston Society of Architects in 1985, a Governor's Design Award in 1986, a Merit Award for landscaping for multifamily housing from the Boston Society of Landscape Architects in 1987, and the Urban Land Institute's 1989 Award for Excellence in the category of Rehabilitation Development.

4. *The Commonwealth Report: Proposals for Capital Improvements, Management Reorganization and Expansion of Resident Services and Opportunities* (Community Planning and Research, Inc., 1979) and *Refinancing and Marketing the Fidelis Way Housing Project* (Walsh & Associates, 1979).

5. A thorough and independent real-time account of this selection process, completed as a Master's thesis in 1982, confirms that the process was thought highly successful by all participating groups at the time (Cunningham, 1982).

6. As of late 1994, there was considerable interest among Commonwealth's tenant leaders in moving the development toward tenant management, though this would presumably still include contracting with a private management company for many day-to-day operations. This potential move, although partly proposed out of frustration with the BHA bureaucracy, seems to stem largely from a hope that formation of a resident management corporation (RMC) could yield to higher per-unit federal subsidies for the development, as has been the case with RMCs elsewhere.

7. The two surveys, conducted in 1979 by Community Planning and Research, Inc., and in 1993 by a team under my own supervision, asked several questions that were phrased and coded identically and that seem to be remarkably comparable samples. Both are stratified by race, ethnicity, and sex and accurately reflect the breakdown of those living at the development at the time; both used adult respondents with a broad range of ages and length of residency in the development. The 1979 sample contained 80 respondents, representing about 30% of the households then living on site, whereas the 1993 sample contained 41 respondents, comprising about 11% of households.

8. In the 1993 survey, residents were asked separately about daytime and nighttime safety; the figures given here represent the average of the two responses.

9. In the late 1970s, approximately 80 of Commonwealth's elderly residents elected to move to a new BHA development for the elderly located nearby; this exodus played a substantial part in the escalating vacancy rate at the development after 1975.

10. The obvious problem of a strategy that focuses scarce dollars on the least disadvantaged of highly disadvantaged places, is that it fails to assist those who live in places that are most in decline. The key to making this an equitable and morally acceptable strategy may be

for housing authorities to acknowledge frankly that there are some places where—due to the density of impoverished families housed together in a devastated neighborhood—the problems are beyond repair through any existing redevelopment program. In these places, untransformable even through the unprecedented generosity of the Boston experiments or the federal Urban Revitalization Demonstration program and unlikely to be able to attract a wider mix of incomes, the best solution may be to rehouse residents elsewhere through voucher programs and, as this is accomplished, to demolish the projects. This kind of triage should, however, be seen as a last resort. To withhold public housing redevelopment dollars from the most distressed neighborhoods will represent yet another blow to the prospects for reinvestment in such places.

REFERENCES

Allston-Brighton Medical Care Coalition. (1978, February 1). *The Commonwealth health improvement program.* Application to the Special Hospital Project Grants Program of the United Way of Massachusetts Bay.

Boston Housing Authority. (1979a, October). *State of the development report.* Boston: Author.

Boston Housing Authority. (1979b, December). *Site selection criteria for substantial rehabilitation. Planning department.* Boston: Author.

Boston Housing Authority. (1981a). *Commonwealth developers kit.* Boston: Author.

Boston Housing Authority. (1981b, December 10). *Commonwealth development: Developer selection results.* Boston: Author.

Boston Housing Authority. (1982, April 16). *Commonwealth: RFP for private management services.* Boston: Author.

Boston Housing Authority. (1993, April). *Tenant demographics report.* Boston: Author.

Bratt, R. (1985, September/October). Controversy and contributions: A public housing critique. *Journal of Housing,* pp. 165-173.

Braverman, J. (1989, June 15). Tenant group loses funds. *Allston Brighton Citizen Item.*

Commonwealth Tenants Association. (1981, October 21). *Tenant evaluation of developer's proposals.* Memo to Pamela Goodman and the Fidelis Way Redevelopment Selection Committee.

Commonwealth Tenants Association Management Committee. (1981, October 7). *Preliminary management recommendations for Fidelis Way.* Memo to Robert Pickette et al., John M. Corcoran Company, & Sandy Henriquez et al., BHA.

Commonwealth Tenants Association, Boston Housing Authority, & Corcoran Management Company. (1983). *Commonwealth management plan: Memorandum of understanding.* Boston: Authors.

Commonwealth Tenants Association & Commonwealth Health Improvement Project. (1980, June 5). *Commonwealth employment project.* Proposal by CTA and CHIP.

Community Planning and Research, Inc. (1979). *The Commonwealth report: Proposals for capital improvements, management reorganization, and expansion of resident services and opportunities.* Report prepared for the Boston Housing Authority and the Commonwealth Tenants Task Force. Boston: Author.

Corcoran, J. M., Sr. (1994). President of John M. Corcoran & Company. Interview with author.

Cunningham, M. O. (1982). *The evaluation of the rehabilitation designs for Fidelis Way.* Unpublished master's in city planning thesis, Department of Urban Studies and Planning, Massachusetts Institute of Technology.

Goodman, P. (1994). Planner and Project Manager of Commonwealth Redevelopment, Boston Housing Authority. Interview with author.

Hanafin, T. (1989, January 7). Fidelis Way: Private managers give new life to public project. *Boston Globe.*

Kahn, R. (1993, October 10). Brighton tenants fear the bad old days. *Boston Globe.*

March, E. (1983). *Money makes it easier: Turning around large troubled housing projects.* Unpublished master's in city planning thesis, Massachusetts Institute of Technology.

National Commission on Severely Distressed Public Housing. (1992, December). *Case studies and site examination reports. Chapter 4: Commonwealth Development.* Washington, DC: Government Printing Office.

Newman, O. (1972). *Defensible space.* New York: Macmillan.

Pickette, R. (1994). President of Corcoran Management Company. Interview with author.

Rapp, D. (1994). Community Organizer and Commonwealth Health and Housing Improvements Program Executive Director, 1979-84. Interview with author.

Reed, J. (1995). Site Manager for Corcoran Management Company, Commonwealth Development. Interview with author.

Seebold, J. (1979, November 29). Task force focus: Fidelis. *Allston Brighton Citizen Item.*

Seebold, J. (1980a, March 3). BHA receiver Spence raises hopes at Fidelis Way. *Allston Brighton Citizen Item.*

Seebold, J. (1980b, June 12). Analysis: Fidelis worked hard to earn funds. *Allston Brighton Citizen Item.*

Trietsch, D. (1994). Allston Brighton Neighborhood Planner for Boston Redevelopment Authority and Cochair of Monastery Hill Planning Task Force, 1979. Interview with author.

Vale, L. J. (1992, May). *Occupancy issues in distressed public housing: An outline of impacts on design, management, and service delivery.* Report prepared for the National Commission on Severely Distressed Public Housing. Washington, DC: NCSDPH.

Vale, L. J. (1993). Beyond the problem projects paradigm: Defining and revitalizing severely distressed public housing. *Housing Policy Debate, 4*(2): 147-174.

Vale, L. J. (1994). Seven kinds of success: Assessing public housing redevelopment efforts in Boston. In W. Preiser, D. Varady, & F. Russell (Eds.), *Future visions of urban public housing: Proceedings of an international forum* (pp. 327-340). Cincinnati, OH: University of Cincinnati.

Vale, L. J. (1995). Transforming public housing: The social and physical redevelopment of Boston's West Broadway development. *Journal of Architectural and Planning Research, 12*(3).

Wallace, Floyd, Ellenzweig, Moore, Inc. (1979, October). *Recommendations of the Monastery Hill Task Force.* Report prepared for the Boston Redevelopment Authority.

Walsh & Associates. (1979). *Refinancing and marketing the Fidelis Way Housing Project.* Boston: Author.

Young, K. (1994). Senior Manager of Corcoran Management Company, former Site Manager of Commonwealth Development (1987-89). Interview with author.

6 Chicago's Mixed-Income New Communities Strategy: The Future Face of Public Housing?

MICHAEL H. SCHILL

Abstract

Public housing in Chicago, Illinois, is virtually synonymous with income and racial segregation. Nevertheless, in the late 1980s, the Chicago Housing Authority obtained authorization from the U.S. Congress to begin an experiment in creating an economically mixed public housing development on the shores of Lake Michigan. The development, Lake Parc Place, opened in 1991. This chapter describes how Lake Parc Place has succeeded in providing tenants with a safe and secure living environment. It also examines some of the limitations of this type of housing development as a model for revitalizing public housing in Chicago and elsewhere in the nation.

For much of its history, the Chicago Housing Authority (CHA) has been embroiled in controversy. The federal takeover of the CHA in May 1995 amid allegations of mismanagement and fraud is only the most recent manifestation of the troubles that have plagued the agency for decades. Years of insufficient operating subsidies, counterproductive federal regulations,

AUTHOR'S NOTE: In addition to the sources listed in the references, the author interviewed representatives of the CHA, HUD, the City of Chicago and Rescorp Realty Corp., as well as tenants of Lake Parc Place and neighborhood residents. Visits to Lake Parc Place took place in 1992 and 1994. Financial support from the Claneil Foundation and the University of Pennsylvania Institute on Law and Economics is greatly appreciated, as is the research assistance of Keith Antonyshyn.

and mismanagement and corruption at the local level have resulted in thousands of housing units that require substantial investment before they can be made habitable. Public housing residents live in overwhelmingly poor and racially segregated areas that can be called "communities" in only the technical sense of the word. Violent crime and drug abuse are ubiquitous; employment and intact nuclear families are the exception.

Prior to the federal takeover and his resignation as Chairman of the CHA, Vincent Lane conceptualized and began implementing an innovative program to ameliorate the legacy of the past. This policy, called the Mixed-Income New Communities Strategy (MINCS), sought to reorient the spending of federal modernization subsidies away from rebuilding the mistakes of the past and toward creating new economically integrated communities. Despite ambitious plans to tear down much of Chicago's high-rise public housing and renovate the remainder to become homes to both the poor and nonpoor, Lane could point to only one completed project, Lake Parc Place, at the time he resigned.

In this chapter, I examine Lane's program to create mixed-income new communities in the place of existing public housing. In particular, I discuss the reconstruction and current operation of Lake Parc Place. Although it is still too early to evaluate its success, data suggest that mixing very poor families with moderate-income working families can create a safe and stable living environment. Nevertheless, the cost of renovating existing public housing developments, the neighborhood conditions surrounding these developments and the difficulties in finding replacement housing for tenants limit the likelihood that such a policy can be replicated on a large scale in Chicago and other American cities.

Public Housing and the Legacy of the Past

The modern public housing program has its roots in efforts by Congress and the Roosevelt administration to stimulate employment in the building trades during the Great Depression. Early initiatives by the federal government to build and operate publicly owned housing were effectively blocked by a court case holding that the federal government could not use its power of eminent domain to assemble parcels of land for housing construction (*U.S. v. Certain Lands in Louisville,* 1935). In a legislative compromise that would have unfortunate consequences, Congress passed and the President signed into law the Housing Act of 1937 that decentralized the program to local governments. Municipalities that wanted to participate in the program could create public housing authorities (PHAs) that would build and retain ownership of public housing. The federal government would provide subsidies for the capital costs of the developments and promulgate regulations to govern their operation.

The decentralized structure of the Public Housing Program contributed directly and indirectly to the concentration of impoverished households in central cities by permitting suburban municipalities to opt out of providing subsidized housing (Schill, 1993; Schill & Wachter, 1995).[1] Poor families wishing to live in public housing were, therefore, frequently unable to obtain accommodations in the suburbs, and, instead, had their choices limited to the cities. In addition, because the bulk of public housing was built on relatively expensive inner-city land, PHAs had to economize on land costs by building at high densities. As these large tower blocks proved to be inappropriate living environments for families with children, households with adequate resources moved elsewhere, leaving in their wake a more intensely poor and marginalized population. In many instances, undercapitalization of public housing resulted in both design and structural flaws. High levels of common facility usage caused frequently substandard building systems to break down. In addition, the inability of parents to monitor effectively their children and the anonymity of the developments led to vandalism and other crimes that, in turn, drove those households who had resources to move away (Newman, 1972).

Federal admission and rent calculation requirements also caused public housing developments to be overwhelmingly composed of extremely low-income families.[2] Although the Housing Act of 1937 reflected Congressional ambivalence with respect to what income groups public housing was to serve, successive statutes increasingly limited admission to the very poor (Schill, 1993). PHAs were required to promulgate income ceilings and evict families whose incomes exceeded these levels. Particularly needy households, such as those displaced by the Urban Renewal Program and in later years the homeless, were given priority in admissions. Current rules require PHAs to ensure that between 75% and 85% of their tenants earn incomes that are less than half of the area median income. In addition, half of all tenants must be selected according to federal priority categories.[3]

In addition to income ceilings, rent calculation rules have caused working-class families to leave public housing in many cities. To assist families with heavy rent burdens, Congress in 1969 enacted the Brooke Amendment that limited rents to one quarter of a household's income.[4] These rent caps created budget deficits for many PHAs and led many to defer maintenance. As buildings began to crumble, those families with sufficient resources moved elsewhere. In addition, in several cities the requirement that tenants pay a set proportion of their incomes for rent created an incentive for families with the highest incomes to leave public housing and obtain less costly housing in the private market (U.S. Dep't of HUD, 1994).

Importantly, local PHAs played an active role in transforming public housing from a utopian ideal to its current dismal reality. Indeed, no public housing authority in the United States has had a more infamous reputation than the CHA. Mismanagement, political interference, and corruption have squandered public resources and caused the deterioration of its stock of public housing (Coffey, 1987; Reardon & Kaplan, 1987). A 1987 report to Congress, reviewing 48 management studies on the CHA issued during the 1980s, found that they "disclosed 798 problems in every critical area of CHA's operations," many of which had existed for years (U.S. GAO, 1989, p. 24). Federal inspections of apartments in nine family public housing developments revealed that 93% had serious physical problems (U.S. GAO, 1989, p. 15). These physical defects included trees growing on roofs, holes in walls and ceilings, broken elevators and inoperative plumbing. A series of internal audits released by the federal government following the May 1995 takeover of the CHA documented widespread mismanagement and procurement irregularities (McRoberts, 1995a).

Inept or corrupt CHA management practices indirectly promoted the concentration of poverty in Chicago's public housing developments by causing households who had resources and options to move away. The concentration and isolation of the poor in Chicago public housing, however, was also intentionally promoted by Chicago's political establishment. Locations for public housing in Chicago were selected in such a way as to reinforce and in some instances, establish racially segregated residential patterns. Following World War II, demand by black households for public housing was particularly strong because of discrimination in the private housing market, the migration of rural blacks to Chicago from the South, and urban renewal projects that bulldozed several neighborhoods with large black populations (Hirsch, 1983; Lemann, 1991). This demand for public housing was met by a large CHA production program, but one in which the sites for new developments were deliberately located in predominantly black neighborhoods.

To ensure that public housing would not be built in white communities, Chicago politicians succeeded in getting the State of Illinois to grant the City Council the right to veto sites proposed by the CHA. A custom developed, by which individual neighborhoods were permitted to reject projects proposed for construction within their borders. As Hirsch describes,

> Of the 33 projects approved between 1950 and the mid-1960s, 25 and a substantial portion of another were located in census tracts containing a black population in excess of 75%. Of the remaining seven developments, six were located in areas undergoing racial transition. By the time the projects were actually completed, only one of the 33 was situated in an area that was less than 84% black; and all but seven of the developments, when actually completed, were located in census tracts that were at least 95% black. (Hirsch, 1983, pp. 242-243)

Chicago's practice of using public housing to promote racial segregation was denounced by a federal court in the landmark case of *Gautreaux v. Chicago Housing Authority.*[5] In its decision, a federal district court held that the CHA had violated the U.S. Constitution by locating public housing in predominantly black neighborhoods and by refusing blacks admission to developments in white communities.

Although some might debate whether the primary cause was the structure of the public housing program, the rules Congress required PHAs to operate under, mismanagement of the program by the CHA, or racially discriminatory siting policies engaged in by Chicago politicians, there can be no question that all these factors have contributed to enormous levels of racial segregation and concentrated poverty in CHA developments.[6] As Table 6.1 indicates, just over one third of all households in the City of Chicago are headed by a black person compared to over four fifths of the households living in Chicago public housing. When only the largest family public housing developments are considered, the proportion of black household heads rises to over nine out of ten.[7] Less than 10% of all households living in Chicago public housing reported earnings from employment in 1991 to 1992, compared to a citywide average of 75%. As might be expected from the low proportion of working households, a large proportion of CHA tenants receive public assistance. For example, among tenants living in family developments, 61% receive Aid for Families with Dependent Children (AFDC).

Although the problems of public housing in Chicago are more severe than those experienced in housing developments in most American cities, in many instances the differences are ones of degree rather than of kind. In many central cities, public housing developments have become home

TABLE 6.1 Characteristics of Chicago Public Housing (1991-1992)

	City of Chicago	*All Chicago Public Housing*	*Family Public Housing Developments**	*Large Public Housing Developments***
Population	2,783,726	86,547	71,456	32,034
Housing units	1,133,039	40,702	29,299	14,320
Vacant units (%)	9.5	15.5	19.6	26.1
White household head (%)	47.9	8.2	1.9	0.2
Black household head (%)	34.9	84.2	92.1	93.2
Hispanic household head (%)	13.8	2.2	1.3	0.1
Households with income from employment (%)	74.8	8.5	10.2	6.9
Households receiving AFDC (%)	14.4***	44.2	61.0	59.8

SOURCE: 1990 Census of Population and Housing Chicago Housing Authority, *Statistical Profile: The Chicago Housing Authority* (1991 to 1992).
* Excludes senior citizen and scattered-site units.
** Includes developments with over 2,000 units.
*** For the City of Chicago, this figure includes recipients of all types of public assistance.

to an overwhelmingly poor and distressed population. This concentration of poverty is attributable to many factors, including federal rules that require PHAs to grant priority in admissions to the most needy.

Chicago's Mixed-Income New Communities Strategy

Ironically, Vincent Lane, a local real estate entrepreneur, took over the CHA in 1988 in the face of federal threats to assume control of the agency on the grounds of mismanagement (Reardon, 1994). In his seven years as Chairman, Lane proposed a number of innovative and sometimes controversial strategies to improve the lives of households living in Chicago public housing. To combat high levels of crime in CHA developments he inaugurated a program of enhanced security and apartment sweeps.[8] To promote economic development, Lane established a program that trained and employed tenants in CHA renovation projects. Perhaps most important, Lane attempted to combat the concentration of poverty that exists in many CHA developments. Toward this end, he articulated and began implementing an initiative called the Mixed-Income New Communities Strategy (MINCS). Under MINCS, mixed-income residential environments would be promoted both by infusing moderate-income tenants into existing public housing developments and by facilitating the dispersal of poor households into relatively higher-income settings.

The MINCS Demonstration

MINCS has two major components: the renovation of existing public housing developments for mixed-income occupancy and the acquisition of replacement housing in privately owned low-density mixed-income developments. To institute MINCS, the CHA lobbied Congress to include in the 1990 Cranston-Gonzalez National Affordable Housing Act a demonstration program open to Chicago and three other PHAs. Congress authorized participating PHAs to lease up to half of the units in selected developments to families who were "low income" but not "very low income." To retain the moderate-income[9] tenants with jobs, rents charged to these tenants would be no higher than statutorily prescribed "ceiling rents." An additional important element of the Chicago plan was that the newly refurbished public housing would be managed by a private management company rather than by the CHA.

The second element of the MINCS demonstration was Congress's authorization for participating PHAs to use their federal operating subsidies to lease apartments for very low-income households in newly constructed or rehabilitated housing owned by private developers. The proportion of units in any privately owned development occupied by very low-income tenants could not exceed 25%. In effect, the private sector units would substitute for the public housing units leased to households earning between 50% and 80% of the area's median income so that there would be no net loss in apartments for the very poor. Participating PHAs were required to screen prospective tenants intensively to ensure that they would be able to participate in the program effectively and not cause problems for their landlords. Those tenants selected to live in the privately owned units would sign contracts with the PHA in which they agreed to remain drug free, not engage in criminal activity, keep their children in school, and participate in a wide variety of social services. Increased income earned by tenants would not be taken into account in computing rents, except if it exceeded 80% of the area median family income. Finally,

PHAs were required to deposit a portion of each rent payment into an escrow account that would be made available to the tenants for the purchase of a home or college tuition upon successful completion of their participation in the program. A tenant could remain a resident of the private housing for seven years, provided that he or she fulfilled the obligations of his or her contract with the PHA.

Lake Parc Place

Following the publication of regulations announcing the MINCS program, Chicago was selected as the only demonstration site.[10] By the time Congress enacted MINCS, the CHA had already begun implementing the first part of the strategy in a public housing development next to Lake Michigan. The project, formerly known as the Victor Olander Homes, was built in 1953 and is located adjacent to Lake Michigan in the Oakland community, less than five miles from the Chicago Loop. Oakland is one of Chicago's most depressed neighborhoods, composed entirely of black persons. As Table 6.2 indicates, from 1980 to 1990, the neighborhood lost over half of its population. Of Chicago's 77 officially designated community areas, Oakland ranks the lowest in median household income (Woodstock Institute, 1994). Its 1990 median household income of $5,128 is less than one fifth the median income for all households in the City of Chicago. Two thirds of all households in the neighborhood earn incomes below the poverty level and almost 60% receive public assistance. Well over half of all persons over the age of 16 are not in the labor force and an additional 19% are unemployed.

The housing characteristics of Oakland also reflect an economically distressed neighborhood. As Table 6.3 indicates, the owner occupancy rate in 1990 was only 5.6%, compared to a citywide average of 41.5%. Over one third of all the housing units in the community were vacant, a six-fold increase from 1980, compared to a vacancy rate of 9.5% for the City of Chicago. In 1992, one in ten buildings in Oakland had been abandoned, compared to fewer than one in 100 for the city as a whole. Median single-family home sale prices, housing values, and apartment rents were also substantially below citywide averages.[11]

The Olander Homes and the Olander Homes Extension were part of a group of six 16-story apartment buildings collectively referred to as the Lakefront Properties. In 1985, the towers had deteriorated to the point where tenants had to be relocated and renovation planned. At the time the

TABLE 6.2 Population Characteristics of Oakland Community Area

	City of Chicago	*Oakland Community Area*
Population 1980	3,005,072	16,748
Population 1990	2,783,726	8,197
Change in population 1980-1990 (%)	−7.4	−51.1
Black population 1990	1,087,711	8,145
Black population 1990 (%)	39.1	99.4
Female-headed households with children 1990 (%)	21.0	30.3
Median income 1980	$15,301	$4,690
Median income 1990	$26,301	$5,128
Change in median income 1980-1990 (%)	71.9	9.4
Families with income below poverty level 1990 (%)	18.3	69.8
Families receiving public assistance 1990 (%)	14.3	58.8
Persons aged 16 and over not in labor force 1990 (%)	36.3	58.4
Persons aged 16 and over unemployed 1990 (%)	7.2	18.7

SOURCES: 1990 Census of Population and Housing.
Chicago Rehab Network, *The Chicago Affordable Housing Fact Book: Visions for Change* (1993).
Woodstock Institute, *Focusing In: Indicators of Economic Change in Chicago's Neighborhoods* (1994).

buildings were vacated, tenants obtained a promise from the CHA that they would be able to return following the refurbishment.

The CHA subsequently obtained $14 million from the federal government to renovate the six buildings. The vacant apartment buildings languished for several years, as controversy surrounded their future. One local developer, with the support of the Kenwood-Oakland Community Organization, proposed demolishing four of the towers and building in their place 3,000 units of privately owned, low-rise housing for low- and moderate-income families. The remaining two towers would have been converted into housing for the elderly. This proposal encountered opposition from HUD, the Mayor of Chicago, and the local alderman. For different reasons, each of these parties objected to the fact that the proposal did not contemplate replacing all demolished housing units with new public housing apartments.[12]

Faced with a federal modernization grant that was insufficient to renovate all six CHA properties and the prospect of losing the money if it

TABLE 6.3 Housing Characteristics of Oakland Community Area

	City of Chicago	*Oakland Community Area*
Housing units 1980	1,174,703	5,210
Housing units 1990	1,133,039	4,335
Change in number of housing units 1980-1990 (%)	–3.6	–16.8
Vacancy rate 1980 (%)	6.8	6.4
Vacancy rate 1990 (%)	9.5	36.4
Change in vacancy rate 1980-1990 (%)	39.2	465.4
Abandoned buildings 1992 (%)	0.8	9.7
Owner-occupied units 1990 (%)	41.5	5.6
Median single family home sales price 1989	$80,000	$17,000
Median house value 1990	$78,700	$50,000
Median rent 1990	$377	$129

SOURCES: 1990 Census of Population and Housing.
Chicago Rehab Network, *The Chicago Affordable Housing Fact Book: Visions for Change* (1993).
Woodstock Institute, *Focusing In: Indicators of Economic Change in Chicago's Neighborhoods* (1994).

did not begin work promptly, the CHA awarded a contract to renovate two of the buildings in 1988. Although at the time the renovation work began, the CHA had planned to admit tenants to the building as it would to any public housing development in the city, these plans changed as Lane assumed control of the agency. The renovated buildings were slated to become the first stage of MINCS. To entice families with employed household heads to move into the buildings, the CHA obtained waivers from HUD's Modest Design Standards that permitted it to include "amenities," such as ceiling fans, miniblinds, closet doors, shower heads, and wood vanity sink fixtures.

In July, 1991, work on the two apartment towers had been completed and they were renamed Lake Parc Place. Each building had 28 one-bedroom units, 85 two-bedroom units, and 28 three-bedroom apartments. The development also included a landscaped play area, security desks, on-site laundry facilities, and a day care center. The cost per unit for the renovations exceeded $70,000 for a total construction cost of close to $20,000,000. The CHA, under the watchful eye of the Lakefront Community Organization, an association representing the interests of the households who were relocated from the six towers in 1985, sought to contact as many relocatees as possible to offer them apartments in the newly constructed dwellings. In addition, the agency advertised to fill the apartments reserved for families with working household heads.

By 1992, Lake Parc Place was almost full. As Lane had expected, the building's location, low rents, and amenities attracted more than enough moderate-income families to occupy one half of the units and achieve the desired economic mix. The median income of the moderate-income tenants attracted to Lake Parc Place was $24,131, compared to $6,196 for the very low-income residents. All the moderate-income families earned income from sources other than public assistance, compared to only one in five of the very low-income families. In terms of household composition, a majority of households in each income group were composed of female household heads with children. Among the very low-income tenants, 96% of all families were either former residents of the Lakefront Properties or their children. Tenants from both income groups were integrated on each floor in the two buildings. Maximum monthly rents for two-bedroom apartments were set at $371, well below prevailing rents for similar apartments.

Mixed-Income Private Housing

A second element of MINCS is CHA's plan to rent 141 apartments for very low-income households in newly constructed or rehabilitated privately owned housing. These units would, in effect, replace the apartments in Lake Parc Place that are currently occupied by households earning between 50% and 80% of the area's median income. In order to achieve the desired one-to-three ratio of subsidized to market rent units, the CHA estimated that 564 units of privately owned housing would have to be renovated or constructed (Chicago Housing Authority, 1992b, p. 9).

To date, no units of privately owned housing have been built or renovated in Oakland under the MINCS plan. Under the original plan approved by HUD, the CHA proposed to lease one quarter of the units to be built by either profit-motivated or nonprofit developers on a parcel of vacant land two blocks away from Lake Parc Place (Chicago Housing Authority, 1992a, p. 9). This land, like much of the land in Oakland, is owned by the City of Chicago, and the CHA's plans contemplated that it would be donated to the CHA. Financial pro formas also assumed that the privately owned housing would receive municipal property tax abatements and federal low-income housing tax credits.

A number of factors have contributed to delay and may curtail the second phase of MINCS. A substantial number of community residents are opposed to the construction of additional low-income housing units in the Oakland neighborhood. This opposition has been voiced by neighborhood block groups and the Kenwood-Oakland Community Organization. In addition, the City Planning Commission recently designated the neighborhood as a Conservation Area. Under state law, the Conservation Plan was developed with the participation of a Conservation Community Council (CCC) appointed by the mayor. Any conveyance by the city to the CHA of land for construction of low- and moderate-income housing would be subject to CCC consultation. Although some members of the CCC are CHA residents, most of them are not. Therefore, controversy in the neighborhood over new housing construction has been reflected in the CCC and has impeded the CHA's plans for acquiring the city-owned land (McRoberts, 1995c).

An alternative proposal for building the "replacement" housing in Oakland involves the demolition of three of the CHA's four remaining vacant Lakefront Property towers. The land on which these towers are built could then be made available for development without having to go through the CCC consultation process. Furthermore, neighborhood opposition to subsidized housing would likely be eased by the demolition of the existing towers. One stumbling block to agreement on this plan is the memorandum of agreement between the CHA and former residents of the Lakefront Properties that guaranteed them the right to return following renovation. Negotiations between the CHA and the Lakefront Community Organization are, at present, ongoing with a compromise expected soon. This plan will likely involve the construction of some units of housing in Oakland and some in other neighborhoods both in the City of Chicago and, perhaps, in the suburbs.[13] Once the land use issues have been resolved, construction may begin, although it is likely that the housing will be built not by private sector investors but by the CHA itself.

Evaluating MINCS

The CHA's MINCS program is motivated by the expectation that mixed-income communities will avoid many of the problems commonly associated with high-density public housing developments. Although it is too early to determine whether either phase of MINCS is a success, preliminary evidence suggests that in many ways the redevelopment of Lake Parc Place has accomplished positive results. Experience with

MINCS also suggests, however, that several of the long-term social objectives of mixed-income public housing may be difficult to achieve.

The Short-Run Experience of Lake Parc Place: Improved Quality of Life

Visitors to Lake Parc Place, particularly those familiar with other public housing developments in the City of Chicago, are immediately struck by the physical quality of the development and its grounds. These include facilities that would be unremarkable in middle-income housing developments but are largely unheard of in public housing. Immaculate laundry rooms, bright and cheerful apartments, secure lobbies, and well-maintained playgrounds continue to set Lake Parc Place apart more than three years after it reached full occupancy.

The social environment at Lake Parc Place is also unlike that which exists in most Chicago public housing. Violent crime is virtually nonexistent. Although some residents have used drugs on the site, swift action by the private management company in charge of the buildings has led to their eviction. Gang activity in and around Lake Parc Place is minimal. Neighborhood complaints are largely limited to unhappiness about littering and occasional loitering on nearby street corners.

A recent survey of tenants living in Lake Parc Place also gives cause for optimism. Rosenbaum and Fishman (1994) report seemingly high levels of satisfaction among residents with life at the development.[14] Over 80% of the tenants responded that they were satisfied or very satisfied with Lake Parc Place. Although higher levels of satisfaction were elicited from very low-income tenants than from moderate-income tenants, almost three in four of the latter group expressed positive feelings about the development. Large majorities of residents said that they were satisfied with building safety, management, and maintenance. The only aspect of the survey that suggests dissatisfaction with Lake Parc Place is the finding that despite a majority responding that they felt the buildings were safe, only 43.4% of the moderate-income respondents felt similarly about the neighborhood. This concern about neighborhood safety was significantly higher than that expressed by the very low-income households living at Lake Parc Place.[15]

The desirability of Lake Parc Place is further demonstrated by the CHA's ability to retain the development's social mix and by its long waiting list. From 1992 to 1994, although the proportion of very low-income tenants inched up to 56%, the CHA has for the most part been able to maintain a social mix in the development. A significant proportion of the households living in Lake Parc Place earn substantial incomes and could afford to move elsewhere. Over 4,000 households are on waiting lists for the very low-income units; 2,000 applicants seek the units reserved for moderate-income families.

The Theory Underlying MINCS

Without minimizing their achievements, the CHA and its former leader, Vincent Lane, had much more ambitious objectives in mind in devising MINCS than just creating a pleasant and safe living environment at Lake Parc Place. The expenditure of relatively large sums of money and careful design guaranteed that Lake Parc Place would be structurally and aesthetically superior to most public housing in the United States. Intensive screening of tenants, careful management of the premises, and vigorous crime prevention activities also made it likely that the buildings would be well maintained and relatively safe. However, Lane also initiated the MINCS program specifically to create a model for how to break down concentrations of poverty in public housing and alleviate the impact social isolation has on the lives of people who are residents of the development. According to Lane,

> Our buildings have become warehouses for the poor. . . . If the Ku Klux Klan had set out to destroy black people, they couldn't have done a better, more systematic job of it than this combination we have of welfare and public housing. We need the right role models to compete with the gangs and drug dealers. (Wagner & Vitullo-Martin, 1994, p. 24)

The concern that poor, predominantly black households are increasingly concentrated and isolated in inner-city neighborhoods has been given widespread attention as a result of William Julius Wilson's influential 1987 book, *The Truly Disadvantaged.* In that volume and subsequent articles, Wilson (1991a, 1991b) argued that concentrated ghetto poverty generates social problems that are different in both magnitude and kind from the problems that afflict poor people in less isolated surroundings. According to Wilson, poor individuals growing up in communities without many employed neighbors are likely to develop weak attachments to the labor force. Lacking opportunities and socialization for employment, ghetto teenagers frequently turn to deviant or illegal activities to earn income and thereby further distance themselves from middle-class norms.

These behaviors and attitudes are then reinforced by peer groups that develop alternative status systems. Getting away with crime or having a child while still a teenager earns admiration; doing well in school or working for modest wages earns scorn (Anderson, 1991; Massey & Denton, 1993). According to Wilson (1991b, p. 12), "the issue is not simply that the underclass or ghetto poor have a marginal position in the labor market similar to that of other disadvantaged groups, it is also that their economic position is uniquely reinforced by their social milieu."

The precise way that living in a neighborhood of concentrated poverty contributes to phenomena such as persistent unemployment, drug abuse, teenage pregnancies, school drop-outs, and welfare dependence remains somewhat mysterious. Jencks and Mayer (1990) suggest two alternative theories. *Contagion* models analogize the behavioral patterns of the ghetto to pathologies that are spread by peer groups, once a critical level of poverty or social isolation has been reached. On the other hand, *collective socialization* theories hypothesize that young people are influenced not so much by peer groups as by the existence or nonexistence of adult role models.

MINCS seems to be based upon a collective socialization model of development. In numerous statements and articles (Johnson, 1991; Lane, 1994, 1995), Lane has emphasized the importance of role models. The idea that contact with relatively higher-income people will improve the "characters" of poor people has a long pedigree in the theory of urban and town planning (Hill, 1875; Sarkissian, 1976). Several recent empirical studies also have shown that a significant relationship exists between the absence of relatively affluent persons in a neighborhood and increased rates of teenage pregnancy, school drop-outs, and welfare dependency (Brooks-Gunn, Duncan, Klevanov, & Sealand, 1993; Clark, 1992; Osterman, 1991).

Although empirical evidence supports the hypothesis that the existence of relatively higher-income people in a neighborhood reduces the likelihood that children will adopt behaviors that distance themselves from middle-class norms, the mechanism by which neighborhood social composition affects behavior is hazy at best. Wilson and Lane explicitly state that working-class neighbors will serve as "role models" for children growing up in their communities. Although references to role models are ubiquitous, the concept is seldom rigorously defined or empirically tested (Jung, 1986), particularly in the context of black youth (Taylor, 1986).

The concept of the role model has its roots in social learning theory. There is much theoretical and empirical evidence to suggest that children learn by imitating the behaviors of other people, particularly when they perceive that the behavior has benefited the model (Bandura & Walters, 1963). Although most of these studies examine discrete activities, such as play and aggressive acts, they also suggest that children will also imitate and internalize more complex behaviors such as altruism (Gruen & Larrieu, 1983, p. 946).

Most often, the social learning literature presumes that the model will be a parent of the child. However, third parties apparently also can be effective models. Very little rigorous work has been done by psychologists to investigate what characteristics a person must possess to become an effective model. Nevertheless, the literature suggests that a person is more likely to be a model if he or she has a nurturant relationship with the child, is someone the child perceives as being powerful, and is somewhat similar either physically or psychologically to the child (Bandura, 1969, pp. 10-11; Cobb, 1992, p. 568; Fischer & Lazerson, 1984, p. 277).

The Prospects for Lake Parc Place Becoming an Effective Mechanism of Social Change

The prerequisites for being a role model are directly relevant to the success or failure of Lake Parc Place. To the extent, as empirical and theoretical evidence seems to suggest, modeling requires a meaningful relationship to exist between the child and the model, merely infusing an impoverished community with working-class families may be insufficient to affect the child's development. Indeed, there is some, albeit very limited, theoretical and empirical basis to be skeptical that the moderate-income families at Lake Parc Place will serve as effective role models to the children growing up in very poor families (Fishman, 1993).

The literature of social psychology is replete with evidence suggesting that it is far from certain that the moderate-income families at Lake Parc Place will meaningfully interact with the very poor. Countless studies and everyday observation reveal that people frequently identify themselves as members of groups based on racial, ethnic, occupational, or class characteristics and seek to differentiate themselves from other outgroups. When members of a group feel insecure about their status, they may seek to draw favorable comparisons and create social distance between their group and subordinate groups (Hewstone, Stroebe, Codol, & Stephenson, 1988, p. 405; Taiffel & Turner, 1986, p. 22).

Geographic proximity among relatively low-status groups, such as that existing in Lake Parc Place, might generate insecurity among members of the higher-status group.[16] Wacquant's (1993) ethnographic account of residents of Chicago's ghetto and the Cité of Paris leads him to conclude that residents tend to establish "microhierarchies," in which working families, "to regain a measure of dignity and reaffirm the legitimacy of their own status in the eyes of society," look with disdain on those who receive public assistance. Similarly, in his ethnography of two predominantly black neighborhoods in Philadelphia, Anderson (1990, p. 159) finds that members of the black middle class, living "within the shadow of the ghetto . . . have great difficulty living close to the black lower classes . . . and gravitate toward those [in the black community] who exhibit a connection with their own position."

Preliminary ethnographic research at Lake Parc Place suggests that the moderate-income tenants may also be seeking to protect their status by differentiating and distancing themselves from the very poor. In 1993, as part of a larger evaluation being conducted by sociologists at Northwestern University, Fishman (1993) conducted in-depth interviews with 20 female adult residents of Lake Parc Place. Her results suggest that both the moderate-income and very low-income tenants at Lake Parc Place are very conscious of distinctions between the two groups. Frequently, residents of Lake Parc Place were characterized as either "working" or "nonworking." As the theory of intergroup relations discussed above might predict, the moderate-income tenants tended to attribute negative characteristics and behaviors, such as laziness, loudness, and messiness to the very low-income residents. Furthermore, there appeared to be extremely limited interaction between members of the two groups. Moderate-income tenants stated either that they were too busy to socialize or that they distrusted or disapproved of their very poor neighbors. These attitudes, together with the fact that turnover rates for the moderate-income households are roughly three times those for the very low-income tenants, suggest that the CHA's stated objective of having the more affluent tenants serve as role models for the very low-income households may be difficult to achieve.[17]

■ Conclusion

It is still too early to evaluate the success or failure of the experiment in socioeconomic mix at Lake Parc Place. The question of whether tenants with jobs can and will serve as models for the very poor residents and their children and thereby affect their aspirations and behaviors can be answered only after several more years of operation. Nevertheless, one thing is clear today: Lake Parc Place is a much more attractive and safe public housing development than it was prior to 1985. It is also superior in almost every respect to other public housing developments in the city. Lake Parc Place has demonstrated that even in economically depressed communities, socially mixed public housing developments can be places where children can grow up in safe and sanitary surroundings without living in constant fear for their lives.[18]

Nevertheless, MINCS is not a panacea for the problems plaguing many public housing developments in the United States. The cost of renovating public housing developments to the standards of Lake Parc Place would be prohibitive. Even if cost were not an issue, it is unlikely that moderate-income households could be lured to many of the most distressed public housing developments, such as the Robert Taylor Homes in Chicago or the Richard Allen Homes in Philadelphia.[19] For residents of these projects, Lake Parc Place may appear to be at best a wasteful use of resources and at worst a mechanism to camouflage their distress. Furthermore, the CHA's difficulty in accomplishing the private construction component of MINCS suggests that transformations, such as the one at Lake Parc Place, will be inhibited as long as the infusion of moderate-income tenants is tied to requirements that public housing authorities replace each unit of public housing that is demolished or rented out to an otherwise ineligible household.[20]

Finally, a careful analysis needs to be undertaken of whether the objective of MINCS, the deconcentration of the inner-city poor, can be achieved more efficiently and effectively through alternative housing policies. The cost of renovating Lake Parc Place (approximately $20 million) was only about $2.5 million less than the amount HUD allocates for new construction of public housing projects, even though the CHA already owned the land on which it was rebuilt. In seeking to remedy the problems posed by concentrated poverty, the CHA has ironically underscored the shortcomings of policies that directly subsidize the suppliers of low-income housing. Although tens of thousands of tenants in Chicago receive no housing subsidies and pay extraordinarily burdensome proportions of their incomes for rent (U.S. Dep't of HUD, 1992), the moderate-income tenants of Lake Parc Place who could afford market rents receive the benefits of federal modernization subsidies and below-market rents. In a supply-oriented housing assistance program, achieving the important objective of economic integration usually requires sacrificing vertical equity. In contrast, in most housing markets, economic integration and an equitable as well as efficient distribution of scarce housing subsidies can be simultaneously achieved through demand-oriented assistance, such as housing vouchers and certificates (Schill, 1993).

NOTES

1. Indeed, even if suburban jurisdictions wanted to build public housing, the Housing Act of 1937 would have made such construction difficult, if not impossible. Under the statute, participating PHAs were required to eliminate one unit of substandard housing for each unit of public housing built. In some jurisdictions, particularly those located in the suburbs, insufficient quantities of slum housing would have made their participation in the program impossible (Wood, 1982, p. 11)

2. For a detailed history of these laws, see Schill (1993).

3. As this chapter goes to press, Congress is considering legislation that would abolish federal preferences and grant PHAs greater discretion in tenant selection.

4. This proportion was subsequently increased to 30%.

5. 296 F. Supp. 907 (N.D. Ill., 1969).

6. Although the levels of racial segregation and concentrated poverty in Chicago public housing are extreme, similar, albeit less severe, patterns exist in other large American cities (Goering, Kamely, & Richardson, 1994; Vale, 1993). Public housing has also contributed to increases in concentrated poverty in Chicago neighborhoods (Massey & Kanaiaupuni, 1993).

7. Racial segregation of blacks in Chicago's public housing is demonstrated in a recent article by Bickford and Massey (1991). The authors computed indices of dissimilarity for public housing residents in 1977 for 45 metropolitan areas and found that the level of segregation in Chicago public housing was the highest. According to Massey, over 92% of all black CHA tenants would have to exchange places with white residents to achieve an even distribution by race. By 1993, the index of dissimilarity value for Chicago had declined only two percentage points (Goering, Kamely, & Richardson, 1994).

8. These sweeps have been suspended by a federal court ruling in *Pratt v. Chicago Housing Authority,* 848 F. Supp. 792 (N.D. Ill., 1994) that held that they constituted warrantless searches without probable cause in contravention of the tenants' Fourth Amendment rights.

9. To distinguish more clearly between the households earning between 50% and 80% of the area's median income and those earning less than 50% of the median, I will characterize the former group as "moderate"-income households and the latter group as "very low" income households.

10. Only one other PHA, the New York City Housing Authority (NYCHA), submitted a MINCS proposal. NYCHA subsequently withdrew its proposal, at least in part because its ceiling rents were sufficient to encourage continued occupancy by moderate-income tenants. The most likely explanation for why other cities did not propose MINCS programs is the difficulty they expected they would encounter in ensuring that the privately owned housing units would be built.

11. Despite its close proximity to the Loop, little economic activity exists in the neighborhood. There are relatively few commercial, industrial, or retail uses in the community. Nevertheless, several important community organizations and service agencies exist in the neighborhood.

12. Some commentators (McCarron, 1988) have suggested that former Mayor Harold Washington and Alderman Timothy Evans opposed demolition of public housing due to fears that their core constituencies would be dispersed.

13. A potential impediment to construction of additional subsidized housing in Oakland is the *Gautreaux* consent decree. Under this agreement, the CHA cannot provide new housing units in areas of the city that have high proportions of racial minorities. Nevertheless, it is likely that the special master administering the CHA's scattered site public housing program will recommend a waiver for the MINCS units on the ground that Oakland is a "revitalizing" community.

14. The survey results are primarily designed to establish benchmark data for future surveys by the authors. The absence of a control group and the fact that these reactions were obtained only a relatively short period after the development was leased makes careful analysis of the tenants' satisfaction levels impossible.

15. Among the very poor tenants, 68.2% of the persons responding to the survey felt that the neighborhood was safe or very safe. The two income groups also differed in their comparisons of the neighborhood surrounding Lake Parc Place to the neighborhoods that they used to live in. Among very low-income households, significantly fewer respondents said that their prior neighborhoods were safe or very safe whereas among the relatively more affluent tenants, a larger proportion felt that their previous neighborhoods were safer than Oakland (Rosenbaum & Fishman, 1994, p. 7).

16. In this sense, MINCS may differ from the Gautreaux-Assisted Housing Program. The program, devised as a remedy in the *Gautreaux* litigation, offers applicants for public housing up to three homes in either the cities or suburbs. Follow-up studies of these households indicate increased levels of neighborhood satisfaction, school achievement, employment rates and social integration (Kaufman & Rosenbaum, 1992; Rosenbaum & Popkin, 1991; Rosenbaum et al., 1991). Unlike the case with Lake Parc Place, however, the putative role models in the Gautreaux demonstration were likely middle-income households who felt no threats to their status from one or two very low-income households moving into their communities.

17. A subsequent survey of tenants of Lake Parc Place (Rosenbaum, Stroh, & Flynn, 1996) indicates that a substantial number of residents engage in volunteer activities and have at least one friend at the development. The study does not, however, indicate whether this interaction occurred between very low-income and moderate-income households or solely within these groups. Interestingly, this study also shows that employment levels among tenants at Lake Parc Place declined over time rather than increasing. Since Lake Parc Place opened in 1991, the cumulative turnover rate for moderate-income households was 31%, compared to 10% for very low-income households. Higher turnover rates for the moderate-income households are predictable, because their greater resources give them the ability to move elsewhere and perhaps purchase homes.

18. For a description of life at a more infamous development, see Kotlowitz (1991).

19. Nevertheless, the CHA has already begun a similar reconstruction project at the notorious Cabrini-Green development (Chicago Housing Authority, 1993). Reports indicate that following its takeover of the CHA, HUD has committed itself to proceeding with a mixed-income strategy at this development (McRoberts, 1995b). Like Lake Parc Place, the

location of Cabrini-Green, within a mile of Chicago's exclusive "Gold Coast" residential area, makes it likely that demand among moderate-income tenants will be sufficient to fill the renovated units.

20. For a critique of Congress's one-for-one replacement requirement, see Schill (1993). Current legislative proposals in Congress would eliminate this statutory requirement.

REFERENCES

Anderson, E. (1990). *Streetwise: Race, class, and change in an urban community.* Chicago: University of Chicago Press.

Anderson, E. (1991). Neighborhood effects on teenage pregnancy. In C. Jencks & P. E. Peterson (Eds.), *The urban underclass* (pp. 375-398). Washington, DC: Brookings Institution.

Bandura, A. (1969). Social learning theory of identificatory processes. In D. A. Goslin (Ed.), *Handbook of socialization theory and research* (pp. 213-262). Chicago: Rand McNally.

Bandura, A., & Walters, R. H. (1963). *Social learning and personality development.* New York: Holt, Rinehart & Winston.

Bickford, A., & Massey, D. S. (1991). Segregation in the second ghetto: Racial and ethnic segregation in American public housing, 1977. *Social Forces, 69*(4): 1011-1036.

Brooks-Gunn, J., Duncan, G. J., Klevanov, P. K., & Sealand, N. (1993). Do neighbors influence child and adolescent development? *American Journal of Sociology, 99*(2): 353-393.

Chicago Housing Authority. (1992a). *Application to the U.S. Department of Housing and Urban Development to conduct a mixed-income new communities strategy (MINCS) demonstration program.* Chicago: Author (mimeo).

Chicago Housing Authority. (1992b). *Statistical profile: The Chicago Housing Authority, 1991 to 1992.* Chicago: Author.

Chicago Housing Authority. (1993). *The urban revitalization demonstration program.* Chicago: Author (mimeo).

Chicago Rehab Network. (1993). *The Chicago affordable housing fact book: Visions for change.* Chicago: Author.

Clark, R. L. (1992). *Neighborhood effects on dropping out of school among teenage boys* (Discussion Paper PSC-DSC-UI-13). Washington, DC: Urban Institute.

Cobb, N. J. (1992). *Adolescence: Continuity, change, and diversity.* Mountain View, IL: Mayfield.

Coffey, R. (1987, May 3). Mismanagement at the CHA? Let's call it "corruption." *Chicago Tribune,* p. 3.

Fischer, K. W., & Lazerson, A. (1984). *Human development: From conception through adolescence.* New York: W. H. Freeman.

Fishman, N. (1993). *Bridging the gap at Lake Parc Place: Social interaction in a mixed-income public housing project* (Master of arts thesis, mimeo). Evanston, IL: Author.

Gautreaux v. Chicago Housing Authority, 296 F. Supp. 907 (N.D. Ill. 1969).

Goering, J., Kamely, A., & Richardson, T. (1994). *Location and racial composition of public housing in the United States.* Washington, DC: U.S. Department of Housing and Urban Development.

Gruen, G. E., & Larrieu, J. (1983). The development of moral values and behavior: Implications for clinical practice. In C. E. Walker & M. C. Roberts (Eds.), *Handbook of clinical child psychology* (pp. 937-957). New York: Wiley.

Hewstone, M., Stroebe, W., Codol, J., & Stephenson, G. M. (1988). *Introduction to social psychology.* Oxford: Basil Blackwell.

Hill, O. (1875). *Homes of the London poor.* London: Macmillan.

Hirsch, A. (1983). *Making the second ghetto: Race and housing in Chicago, 1940-1960.* Cambridge, UK: Cambridge University Press.

Jencks, C., & Mayer, S. E. (1990). The social consequences of growing up in a poor neighborhood. In L. E. Lynn, Jr., & M. G. H. McGeary (Eds.), *Inner-city poverty in the United States* (pp. 111-186). Washington, DC: National Academy Press.

Johnson, S. (1991, August 15). Lake Parc Place: New look, new tenants, new face for CHA. *Chicago Tribune,* p. 1.

Jung, J. (1986). How useful is the concept of role model? *Journal of Social Behavior and Personality, 1*(4): 525-536.

Kaufman, J. E., & Rosenbaum, J. E. (1992). The education and employment of low-income black adults in middle-class white suburbs. *Education Evaluation and Policy Analysis, 14*(3): 229-240.

Kotlowitz, A. (1991). *There are no children here: The story of two boys growing up in the other America.* New York: Anchor.

Lane, V. (1994, January 7). High-rises part of strategy in housing. *Chicago Tribune,* p. 22.

Lane, V. (1995). Best management practices in U.S. public housing. *Housing Policy Debate, 6*(4): 867-904.

Lemann, N. (1991). *The promised land: The great black migration and how it changed America.* New York: Knopf.

Massey, D. S., & Denton, N. A. (1993). *American apartheid: Segregation and the making of the underclass.* Cambridge, MA: Harvard University Press.

Massey, D. S., & Kanaiaupuni, S. M. (1993). Public housing and the concentration of poverty. *Social Science Quarterly, 74*(1): 109-122.

McCarron, J. (1988, August 28). "Reform" takes costly toll. *Chicago Tribune,* p. 1.

McRoberts, F. (1995a, June 10). CHA audits show house of horrors; $26 million allegedly was stolen, wasted. *Chicago Tribune,* p. 1.

McRoberts, F. (1995b, July 15). HUD would split new Cabrini between poor, working poor. *Chicago Tribune,* p. 5.

McRoberts, F. (1995c, November 16). Kenwood, Oakland reject CHA plan; property values among concerns. *Chicago Tribune,* p. 1.

Newman, O. (1972). *Defensible space.* New York: Macmillan.

Osterman, P. (1991). Welfare participation in a full employment economy: The impact of neighborhood. *Social Problems, 38*(4): 475-491.

Pratt v. Chicago Housing Authority, 848 F. Supp. 792 (N.D. Ill. 1994).

Reardon, P. (1994, June 24). For six years under Lane, CHA didn't have "scandal-plagued" before name. *Chicago Tribune,* p. 16.

Reardon, P., & Kaplan, J. (1987, February 8). CHA house cleaning a curious tradition; law agencies get call for dirty work. *Chicago Tribune,* p. 1.

Rosenbaum, J. E., & Fishman, N. (1994). *The early reactions of Lake Parc Place residents.* Evanston, IL: Author (mimeo).

Rosenbaum, J. E., & Popkin, S. J. (1991). Employment and earnings of low-income blacks who move to middle-class suburbs. In C. Jencks & P. E. Peterson (Eds.), *The urban underclass* (pp. 342-356). Washington, DC: Brookings Institution.

Rosenbaum, J. E., et al. (1991). Social integration of low-income black adults in middle-class white suburbs. *Social Problems, 38*(4): 448-461.

Rosenbaum, J. E., Stroh, L., & Flynn, C. (1996). *Lake Parc Place: A study of the first four years of a mixed-income housing program.* Evanston, IL: Northwestern University Center for Urban Affairs.

Sarkissian, W. (1976). The idea of social mix in town planning: An historical review. *Urban Studies, 13*(3): 231-246.

Schill, M. H. (1993). Distressed public housing: Where do we go from here? *University of Chicago Law Review, 60*(2): 497-554.

Schill, M. H., & Wachter, S. M. (1995). The spatial bias of federal housing law and policy: Concentrated poverty in urban America. *University of Pennsylvania Law Review, 143*(5), 1285-1342.

Taiffel, H., & Turner, J. C. (1986). The social identity theory of intergroup behavior. In S. Worchell & W. G. Austin (Eds.), *Psychology of intergroup relations* (pp. 7-24). Chicago: Nelson Hall.

Taylor, R. L. (1986). Black youth and psychosocial development: A conceptual framework. In R. Staples (Ed.), *The black family* (pp. 201-210). Belmont, CA: Wadsworth.

United States v. Certain Lands in Louisville, 78 F.2d 684 (7th Cir. 1935).

United States Department of Housing and Urban Development. (1992). *The location of worst case needs in the late 1980s: A report to Congress.* Washington, DC: Government Printing Office.

United States Department of Housing and Urban Development. (1994). Explanation and justification for the Housing Choice and Community Investment Act of 1994. *Congressional Record* 140: S4838.

United States General Accounting Office. (1989). *Public housing: Chicago Housing Authority taking steps to address longstanding problems.* Washington, DC: Government Printing Office.

Vale, L. J. (1993). Beyond the problem projects paradigm: Defining and revitalizing "severely distressed" public housing. *Housing Policy Debate, 4*(2): 147-174.

Wacquant, L. J. D. (1993). Urban outcasts: Stigma and division in the black American ghetto and the French urban periphery. *International Journal of Urban and Regional Research, 17*(3): 366-383.

Wagner, R. F., Jr., & Vitullo-Martin, J. (1994). Revival of public housing. *City Journal, 4*(2): 21-30.

Wilson, W. J. (1987). *The truly disadvantaged: The inner city, the underclass, and public policy.* Chicago: University of Chicago Press.

Wilson, W. J. (1991a). Public policy research and *The Truly Disadvantaged.* In C. Jencks & P. E. Peterson (Eds.), *The Urban Underclass* (pp. 460-481). Washington, DC: Brookings Institution.

Wilson, W. J. (1991b). Studying inner-city social dislocations: The challenge of public agenda research. *American Sociological Review, 56*(1): 1-14.

Wood, E. (1982). *The beautiful beginnings, the failure to learn: Fifty years of public housing in America.* Washington, DC: National Center for Housing Management.

Woodstock Institute. (1994). *Focusing in: Indicators of economic change in Chicago's neighborhoods.* Chicago: Author.

7 Return From Abandonment: The Tenant Interim Lease Program and the Development of Low-Income Cooperatives in New York City's Most Neglected Neighborhoods

ANDREW WHITE
SUSAN SAEGERT

Abstract

During the 1970s, the New York City government began to take control of apartment buildings whose landlords had failed to pay property taxes. This city-owned housing stock came to present an enormous potential resource. Advocates soon joined with officials to craft experimental programs that offered ownership of the occupied apartment buildings to private landlords, community groups, and tenants. While the other programs have since been dropped or thoroughly revised, the "Tenant Interim Lease" (TIL) program continues as one of the primary methods of privatizing city-owned property and has been responsible for converting 527 buildings with 11,620 apartments into tenant-owned, low-income cooperatives. The program provides an innovative model for training tenants in low-income neighborhoods and has the added benefit of involving the residents in stabilizing their communities.

■

A decade ago in the dead of winter, the tenants of 29-33 Convent Avenue in West Harlem routinely went without heat or hot water for weeks at a time. Their landlord had not paid property taxes in years and no longer maintained the building, although he continued to collect rent from those few tenants so scared of being evicted that they felt compelled to pay. Drug dealers lived on almost every floor; the front door did not lock, the intercom did not work, and if it was raining outside, then it was raining on the sixth floor.

Their story is typical of many low-income tenants in Harlem and across the city, where thousands of buildings have been abandoned by landlords unwilling or unable to provide services, keep up with their mortgages, and pay their property taxes. From the 1970s until 1993, New York City took ownership of more than 7,500 occupied apartment buildings, whose owners, like the real estate speculator who owned 29-33 Convent Avenue, had fallen far behind in their tax payments. In this way, the government itself became the largest residential landlord in the poorest parts of the city and, by the late 1980s, owned more than half of the rental housing stock in Harlem.

City ownership has not always turned out to be a successful intervention strategy. Much of this housing is dilapidated and many buildings have languished in the hands of government managers for as long as a decade or more before being rehabilitated and sold to private buyers. In the 1980s, it became routine for newspapers to carry stories about tenants freezing to death in city-owned buildings with inoperable boilers and structural deterioration.

Yet in many cases, city ownership has led to the creation of stable, community- and tenant-controlled housing, thanks to innovative programs developed over more than two decades by housing activists and researchers, tenant groups, and city policy makers. For example, the city Department of Housing Preservation and Development (HPD) took title to 29-33 Convent Avenue in the summer of 1987, and although the tenants there continued to struggle for a number of difficult years with inadequate heat and hot water and other problems, they have now gained the quality housing and localized control they sought for so long. Today, the drug dealers are gone, the roof and boiler have been replaced as have the electrical and plumbing systems, the apartments have new bathrooms and kitchens as well as parquet and linoleum floors, and workers are currently repairing the fire escapes and the ground-level cement and masonry work. The rehabilitation was overseen by the tenants in collaboration with HPD under the auspices of the Tenant Interim Lease (TIL) program and paid for through HPD's capital budget housing program. As soon as work is completed, the tenants will buy the 47-unit building.

About a third of these tenants are on public assistance or disability, and many are elderly and retired. Most of the rest fall in the category of the working poor. Yet, like hundreds of other New York City tenant associations before them, they are creating a limited equity cooperative that will provide affordable housing for at least a generation.

Abandonment and Renewal

Low-income New Yorkers have long faced a severe shortage of affordable housing. In the 1970s, the city lost hundreds of thousands of units of low-cost housing, as landlords abandoned their properties. Since the 1980s, more than 5,000 homeless families at a time have crowded into shelters and welfare hotels, with hundreds forced to sleep each night on the floors and chairs of emergency assistance offices. During the five-year period ending in 1992, nearly a quarter of a million New Yorkers spent time in a homeless shelter, including one of every 20 city children and one of every 12 African American New Yorkers (Culhane et al., 1994).

The thousands of buildings taken over by the city government in lieu of unpaid taxes (a status known as *in rem*) became a focus of housing preservation efforts during three consecutive mayoral administrations from the 1970s to today. While bureaucratic stasis and tight budgets left many buildings in unsatisfactory condition for long periods, the city also pursued programs that sold rehabilitated in rem properties to community nonprofits and tenants. Other programs have turned buildings over to private landlords.

The TIL program has proven especially successful in providing decent and relatively safe housing for the city's poorest residents (68.5% of the residents in TIL buildings have incomes below the poverty level). Since its inception in 1978, TIL has been responsible for the creation of 529 tenant-owned cooperatives with a total of 12,922 apartments. Another 210 buildings with 4,128 apartments are currently in the transitional phase of the program on their way toward cooperative ownership, with more buildings entering every month (figures as of March, 1996).

These tenant-owned cooperatives provide more than a long-term affordable housing resource. Limited equity cooperatives hold intrinsic value in the civic life of low-income communities, because they catalyze organizing efforts and promote high rates of community and electoral participation. The program's successes and failures have also shed light on the process of cooperative development among low-income tenants, including the tenants' training and education needs and the types of tools and technical assistance that any tenant-ownership or tenant-management program must have in order to succeed.

The TIL program makes policy sense especially in a context of high private sector abandonment, a shortage of affordable housing resources, and low political involvement among residents of low-income communities. However, TIL is not an easy sell to government officials who tend to have a top-down style of policy creation and implementation, for the program owes its success as much to the tenants as to the bureaucrats. The New York experience has shown that for TIL to be successful, officials must be willing to turn a high degree of power over to tenants, while still providing substantial rehabilitation and operating subsidies. As a result, the program has not been a centerpiece of city housing policy. Whether the mayor is Republican or Democrat, the program usually moves forward only because of active support at the middle and lower echelons of the HPD hierarchy. Ultimately, the key to TIL's survival has been the grassroots enthusiasm of the tenants themselves.

■ Background

New York's civic culture has had a high awareness of the need for low-income housing since at least the turn of the century. The city's unparalleled stock of public housing is evidence enough, with its population of about 600,000 residents. Yet, this awareness has failed to compensate for unyielding poverty and the inability of the private housing market to supply adequate apartments in poor neighborhoods.

New York's current housing crisis has its roots in the vicious cycles of disinvestment that hit large parts of the city in the 1960s and 1970s. By the mid-1970s, as New York's government foundered in the midst of a crippling fiscal crisis, landlord abandonment had become endemic. Officials and activists began searching for ways to stop the disastrous destruction of already scarce housing resources. Vacancy rates for all rental housing were well below 3%, and apartments were disappearing at a rate of about 31,000 units per year (Bach & West, 1993, pp. 13-15). With the government in desperate need of increased revenue and activists clamoring for a housing rescue operation, the city began to take title to properties whose owners had failed to pay their taxes for more than a year. If government could intervene early enough, the policy makers believed, the buildings could be stabilized before they reached the point of no return.

At first, this approach was undermined by the city's habit of selling these in rem buildings at public auction before any repairs had been made. Studies quickly found that most of the auctioned buildings never returned to the tax rolls because of their decrepit condition and low-income tenancy. Many were soon back in city ownership (Laven, 1978, p. 20).

By 1981, the government owned 3,926 inhabited buildings with a total of 37,341 mostly occupied apartments (Cohn, 1981, p. 3). The tenants in these buildings were among the city's poorest residents, those who could least afford to pay rents that would make ownership profitable to private landlords. In addition, the city owned vacant lots and tens of thousands of units in empty buildings.

Some years before, grassroots community organizations throughout the city had begun to gain control of a small number of apartment buildings whose landlords had given up hope of paying their taxes or mortgage debts. On Manhattan's Lower East Side and a few other sections of the city, sweat-equity homesteading and federally financed rehabilitation work began in earnest. In 1978, a coalition of housing activists and a handful of members of the City Council issued a report, denouncing the auction program and proposing an array of reforms that were to serve as the blueprint for a new infrastructure promoting community and tenant ownership of tax-foreclosed housing. City bureaucrats soon joined with activists to devise a series of programs that, in the ensuing years, brought tenants, community leaders, private contractors, philanthropists, and city government together in a sometimes fractious partnership to stabilize successive waves of foreclosed housing (Sierra, 1993).

During the same period, tenants, encouraged by the Harlem rent strikes of the late 1960s, organized frequent rent strikes against landlords in deteriorating neighborhoods. As the landlords withdrew services, an increasing number of tenant associations succeeded in having the courts appoint independent building administrators through a state law established to give tenants an alternative to landlord negligence. As a consequence, tenants often became closely involved in the management of many low-income buildings, even though they did not officially own the property.

Inspired in part by this wave of tenant self-sufficiency, the TIL program became part of the assortment of program ideas put forward by activists and policy makers. TIL was an offshoot of an earlier, unsuccessful direct sales program intended to transfer ownership to tenants but that had floundered because of overwhelming legal and regulatory requirements. The paperwork was streamlined, model bylaws for tenant associations drawn up, and in the first two years of TIL's existence, 1978 and 1979, about 20 buildings a month were accepted into the program. Residents were given control of their rent revenues and began refining their property management skills. Many of them moved to full tenant ownership within five years.

Throughout the 1980s, the city continued to take title to a succession of properties abandoned by their landlords. HPD soon became notorious as the largest slumlord in town, and residents in city-owned tenements suffered conditions as bad or worse than any privately owned slum building anywhere in the city. Even as advocates called for expanding the size and scope of TIL and other alternative management programs, the housing department struggled to find money to maintain some semblance of service to its tenants.

Finally in 1987, the administration of then-Mayor Edward I. Koch proposed a 10-year, $5 billion housing plan funded by revenues from bonds floated by the city and state governments, as well as federal Community Development Block Grants, Low-Income Housing Tax Credits, and various specialized HUD programs for the elderly and disabled. The plan placed a high premium on the development of new buildings, including middle-income and one- and two-family homes; but it also set the ambitious goal of rehabilitating nearly 170,000 apartments citywide—including most of the tax-foreclosed in rem stock. With maintenance and operations of in rem housing, costing city taxpayers between $100 million and $200 million a year, the administration became increasingly intent on getting this inventory out of its hands.

Late in the Koch years, a heavier emphasis was placed on selling city-owned buildings to for-profit private landlords, many of them huge management companies based in the New York suburbs. Officials claimed this was a more efficient method than developing tenant or community-based management. But this sales program also had its downside, with extensive reports of harassment, fraud, and unwarranted evictions by landlords. During the first year of the administration of Mayor David Dinkins, the for-profit sales program was phased out and the city's commitment to TIL leapt to about $28 million annually. The money was dedicated primarily to rehabilitating the buildings, while they were managed by tenants in preparation for purchase from the city. By this time, the city's much broader investment of hundreds of millions of dollars a year in housing rehabilitation and development had begun to lift many of the most neglected neighborhoods out of the doldrums, thus helping to stabilize communities where many TIL buildings are located.

In recent years, the city has invested more in housing than the next 50 largest United States cities combined (Dinkins, 1993). The effort has brought relative stability to scores of once forsaken blocks in the South Bronx and Manhattan's Lower East Side, and to a lesser extent, the same is true in sections of Harlem and the aging square miles of central and eastern Brooklyn. Poverty and economic disinvestment continue to take a heavy toll, however. In 1992, almost a quarter (24.4%) of the households in New York reported incomes below the federal poverty level, an increase of more than 3% in two years. Nearly one of every four New York households received some form of public assistance. The Census Bureau defines more than 200,000 New York households as crowded or severely crowded, primarily in low-income neighborhoods (Stegman, 1993). Meanwhile, rents at the bottom of the income scale are rising far faster than those at the top (New York City Rent Guidelines Board, 1994). In 1993, vacancy rates for apartments that would be covered by the AFDC rent allowance fell to such a low level that the percentage could not be reliably estimated. Vacancy rates of under 2% were reported for units renting for less than $500 per month (Blackburn, 1995).

Even so, the city ceased taking ownership of tax delinquent housing in 1993, and much of it is rapidly deteriorating. Analysts say another 20,000 units are on the verge of foreclosure and abandonment (White, 1995). The administration of Rudolph Giuliani has put forward a new set of programs intended to help private landlords work out tax and mortgage debt problems and retain ownership of distressed apartment buildings by obtaining city rehabilitation subsidies. The mayor has promised that in cases where the work-out plans fail and the landlords do not cooperate, tenants will be offered the opportunity to become owners through a program derivative of TIL, though at this point details are sketchy. In the meantime, the TIL program is moving ahead with hundreds of the buildings still in city ownership. Congress's late-1995 elimination of new Section 8 rent subsidies, however, which provide long-term operating support for many co-ops, has left the future scale of the program in doubt.

How TIL Works

First and foremost, tenant management is what TIL is all about. The name, Tenant Interim Lease, refers to the period between the moment tenants gain city government approval to join the program and the time they purchase the building and convert it into a low-income cooperative.

During that interim, the tenant leadership holds a lease on the building and is responsible for collecting rent, selecting new tenants to fill empty apartments, soliciting bids for repair and maintenance work, paying contractors, acting as the liaison between all of the tenants and HPD—and in many cases, pursuing legal action in housing court against residents who fail to pay their rent.

To get into the program, tenants have to get organized. At least 70% of the apartments in the building must be occupied by tenants recognized by HPD as "legal" occupants (the city has been battling squatters for control of a number of apartments for several years). The tenants have to form an association in which more than half of the occupied apartments are represented, hold meetings, and elect a group of officers. And when they apply to the program, the city has to verify that there are no other competing claims for the property, such as a former landlord seeking to redeem the building by paying off back taxes, and no prior commitments by the city to sell or demolish it. Finally, before the application can be accepted, the tenants have to take part in two introductory classes, and the association officers must attend a series of classes on building management, financial management, and maintenance and repair, all taught by the Urban Homesteading Assistance Board (UHAB), a nonprofit organization founded in the mid-1970s.

Once the city is convinced that the tenants are solidly organized and they are accepted into the program, HPD staff assesses the condition of all the building systems and determines if there are any structural flaws that need immediate attention. They meet with tenants to discuss what sort of improvements are necessary to comply with housing codes, and if a complete rehabilitation is planned, they begin working with tenants to draw up the plans. For most of the 1980s, such comprehensive work was rare; TIL buildings received a new boiler and roof, if it was needed, as well as an overhaul of the electrical system and sometimes a new plumbing system. But only if the tenants were particularly combative could they win more than $5,000 worth of work per apartment. These numbers have changed considerably during recent years. Under the Dinkins administration, investment reached as high as $15,000 or $20,000 per apartment in the most dilapidated buildings—partly because these properties had been in city ownership longer and were in worse condition (Peirce & Guskind, 1993, p. 35). Today, the Giuliani administration is spending as much as $50,000 per apartment in each of its three housing privatization programs, including TIL. Because of a backlog, however, buildings in TIL must now wait up to three years after entering the program to receive this investment.

Upon entering the program, the tenant association has to raise rents to at least $55 per room. Most city-owned buildings have not had even one rent increase in the years since their landlords abandoned them, so even these low TIL rents can represent a large increase. Still, the change brings revenues closer in line with the expenses of management and routine maintenance, and it gives tenants an inkling of what it will take to run the building as a cooperative. It also saves the city money and underlines one of the most direct benefits of the program; according to HPD, it costs the city an average of about $500 a year to manage an apartment in the TIL program, above and beyond the rents the city collects, as compared to $2,000 a year for an apartment in central management (Peirce & Guskind, 1993).

Nonetheless, the city has often resisted tenants' efforts to get into the program (Leavitt & Saegert, 1990, pp. 69-72). In some cases, officials have simply told tenants that the program was not accepting any more applications. The reasons are unclear, though the HPD bureaucracy has long been under the gun from outside fiscal monitors to keep a tight rein on agency operating expenditures, a task that many officials may consider too difficult under the decentralized tenant management program. In any case, this hurdle is just one sign of the stubbornness required of tenants to make it into and through the program, maneuvering the bureaucracy and at times seeking the support of local political leaders before finally being allowed into TIL.

Under the privatization-friendly administration of Mayor Giuliani, tenants have had to organize themselves to move their buildings into the TIL program rather than one of the alternative sales programs, including the Neighborhood Entrepreneurs Program (NEP), a new program for small, for-profit landlords, and the Neighborhood Revitalization Program (NRP) that turns buildings over to community groups for rehabilitation with federal Low-Income Housing Tax Credits. The administration has shifted the majority of HPD capital funds into improvement of the tax-foreclosed stock in preparation for sale to private owners, including tenants—$20 million for TIL, $49 million for NEP, and $39 million for NRP—in the budget for fiscal year 1996.

In TIL, tenants pay $250 to buy their apartment (a price that has not changed since 1979) and clauses in the sales documents allow shareholders to sell at only a slight profit and only to people of limited income. For tenants, however, there is no point in purchasing the building until all the repairs have been completed and done well. Otherwise, the cooperative shareholders will have to pay for the work themselves.

The low quality of the workmanship of government contractors is the single most consistent complaint of tenants in the TIL program. It is theoretically up to the tenants to bid out all except the largest projects; for repairs costing more than $2,000, the tenant managers must receive at least three bids and choose the lowest. But increasingly over the years, rehabilitation work has been contracted and overseen by city staff. It is on the largest jobs overseen by the city that many of the horror stories take place.

"I have had contractors come into a twenty-unit building and leave two bathrooms intact for months at a time," says Bruce Morris, a UHAB coordinator. "Another building, they took out all the kitchens." All too often, he explains, contractors approach a mostly occupied building as if it were empty. "It may fit their schedule, but it creates all kinds of problems." Many tenants in the program report similar situations.

Although the rehabilitation period is stressful, it is also a time when tenants are likely to express their will most forcefully and exercise their newfound power. At 29-33 Convent Avenue, for example, Thia Calloway and her fellow tenants used some of their rent money to hire an outside expert, a plumbing consultant, for $600 a week to double-check the work of the general contractor every day. "They wanted to laminate over crumbling walls," Calloway says. "We said no way, demolish these and put up new ones." The contractors did as they were told. Calloway recalls,

> It was important to have someone to deal with this every day, and to take complaints from other tenants. I was about to have a nervous breakdown when all of this started. I would sneak out because I did not want any more tenants yelling at me. They had to let out their frustration and anger. They were living out of boxes during the construction work, and using tiny temporary sinks.

This dissension can sometimes lead to a total breakdown of cooperation within the building and occasionally to a return to HPD control and removal from TIL. More commonly, tenant micro-management can chafe with the interests of city bureaucrats. As a result, HPD has over the years pulled back some of the responsibility from the tenants for oversight and management. To some degree, the changes have weakened buildings' experience of independence and self-reliance, originally the core training elements of the TIL program. Now, resident management in TIL is not an entirely realistic experience of what it will be like once they are on their own.

Inevitably, tension between government staff and tenants has led to dissension and delays in moving buildings through the program. The current administration at HPD is moving closer toward resolving some of these issues by standardizing the scope of work that every TIL building can expect before leaving the program. Officials have also expressed interest in clarifying the city's contract with the tenants, establishing a definite time period within which HPD will bring the buildings up to a certain standard, arrange for long-term subsidies, such as Section 8, if they are needed (and available) and provide training through UHAB. In return, the tenants will have to meet certain management goals within a reasonable time frame, with the implicit threat that if the goals are not met, the city may take the building back and find another buyer.

■ Organizing for Ownership

Two surveys of 5,700 residents of currently or formerly in rem buildings have shown that the vast majority would like to see their buildings become tenant co-ops (Saegert, 1993; Task Force on City Owned Property, 1996). However, the road from desire to actual ownership is long and difficult.

Tenants face many challenges, both internal and external to the building, that must be overcome after taking control of the building and before its purchase. The difficult process of developing an effective tenants' association is exacerbated when leaders and active residents are stymied by bureaucratic unresponsiveness, long waits for promised improvements, and frequent changes in policies and operating procedures brought about by shifting city politics and the pressures for fiscal austerity. They must rely on the strength of their commitments to remain in what are often almost uninhabitable circumstances and to work together in communities, in which distrust and negative expectations about the future are the norm. When landlords withdraw services and the city takes ownership, tenants often see few alternatives to self-ownership. They can stay for the time being under the control of HPD's Division of Property Management, where an overextended and underfunded maintenance and administrative staff attempt, often in vain, to maintain some semblance of housing standards. Or they can wait until their building is sold to a nonprofit community group or a for-profit landlord, alternatives that many residents are unaware exist and that few would choose. Options for moving elsewhere are few, given the tight housing market and the 230,000 households on the public housing and Section 8 waiting list.

In early examples of tenant self-management (Leavitt & Saegert, 1990), tenants often drew on long-standing social ties formed while their buildings were landlord-owned and were often strengthened by a collective fight against deteriorating services. Tenants depended on these relationships first to help each other survive and then as a basis of organizing. Older people, especially older women, emerged as leaders, mobilizing their friends and acquaintances to save their homes. For many of the early leaders, relocation would have meant giving up the social and physical worlds that had taken a lifetime to build.

Gender also affected responses to abandonment. Female-headed households are prevalent in city-owned buildings. Their style of organizing is often built on the social ties that had helped them provide homes for their families on too little income. Male leaders more often saw their work in buildings as an extension of their jobs or community leadership roles (Leavitt & Saegert, 1990).

As tenant ownership became more widespread, the length and nature of tenant social networks and commitment to place varied according to the history of the community. In Harlem, the founders of many of the first co-ops clung to their homes and community in defense of their memories and hopes for Harlem as a historically significant site in the history of African Americans, a history in which their own lives had been embedded. Later, the process of organizing for ownership became the glue that brought people together. As Thia Calloway, the vice president of the tenant association of 29-33 Convent Avenue and president for seven grueling earlier years, stated,

> People who never talked to each other before are talking all the time. We have a wonderful mix of people, by race, age and work. Whites, blacks, Hispanics. I'm the token Asian. There are some tenants who never finished high school and one who has a Ph.D. But we have all learned a great deal. I have seen individuals develop more confidence in their capabilities, develop leadership qualities. We have all learned to grow. We have all had to stretch ourselves to make this happen.

The interplay of internal and external dynamics has changed as communities, city policy, and the broader social context changed. The earliest organized tenant associations were pioneers, aided mainly by advocacy groups and sympathetic politicians. Most of the first self-managed buildings paid for all their own repairs and improvements. Their adversarial relationships with the city led them to try to take control as quickly as possible, a goal often frustrated by inexplicable delays and setbacks. These tenants also often had trouble getting the subsidies to which they were entitled because of bureaucratic bottlenecks. Tenant associations that formed later benefited more from the TIL program both in terms of repairs and subsidies.

Buildings that entered TIL later encountered a better organized program with more resources to provide. However, their own internal resources were often more strained. The later organized buildings had shorter organizational histories and, thus, often less leadership sophistication. They also formed their tenant associations at a time when there was little popular political activism aimed at improving conditions for the urban poor and minorities. Thus, there were fewer technical assistance groups around to support and advise them. For example, legal aid lawyers were very often key sources of assistance and support for the first tenant co-ops. Their presence in poor communities has been drastically reduced during the last dozen years. Over the years, UHAB's contract with the city also became more restrictive, eliminating payments for organizing visits prior to a building's entry into the TIL program and closely defining what visits were covered by the contract.

Meanwhile, tenants faced more desperate conditions, in that their buildings had often deteriorated further and the specter of homelessness was more evident in their communities. The recurring weakness of the local economy led to high levels of unemployment. Many of the later organized buildings included significant numbers of tenants with a variety of disabilities. This combination of factors makes it less likely that tenants could have achieved self-management without a relatively strong TIL program. These conditions also strained the informal helping relationships that were at the core of collective efforts to take control of buildings.

The degree of connection among residents and with the neighborhood is related to the general stability or turmoil the community has experienced. For example, the length of individual tenancies in buildings owned by the city averaged under five years in the Bronx but nine years in Brooklyn. Bronx residents also reported more interest in moving and less involvement with community organizations and institutions (Saegert, 1993; Task Force on City-Owned Property, 1996).

Successes

Despite the difficult economic context and worsening plight of low-income housing in New York, the TIL program has achieved a surprising level of success. It has produced nearly 12,000 units of housing for low- and moderate-income New Yorkers, and most of that housing has been decent, safe, and affordable. In the surveys of city-owned and formerly city-owned buildings in the Bronx and Brooklyn, tenant-owned co-ops scored much higher on management quality and the provision of basic services, like heat and hot water, than did buildings owned by the city or sold to private landlords or community groups. Security was consistently better in tenant-owned co-ops and almost all the surveyed buildings in which tenants said crime and drugs were not a problem were tenant-owned co-ops.

Furthermore, the co-ops retained a high level of affordability despite the better services provided. In the Bronx, over 40% of the co-op residents reported incomes under $10,000 a year, and over 80% had incomes under $20,000. In Brooklyn, incomes were slightly higher but comparable, and the average per-unit rent was slightly more than $300 per month (Saegert, 1993; Task Force on City-Owned Property, 1996).

The process of becoming a tenant owned co-op requires significant tenant participation. Survey data indicate that such participation persists after sale to tenants. Compared to their peers in other types of current and former city-owned buildings, co-op residents reported the highest level of attendance at tenant meetings, the highest levels of leadership activities, and the greatest amount of informal socializing and assistance.

Tenant organizing for self-ownership has also contributed to residential stability, with co-op residents in the Bronx having lived in their buildings, on average, from six to seven years, and in Brooklyn, over 13 years (Saegert, 1993; Task Force on City-Owned Property, 1996). The stabilizing effect of co-op ownership can be seen, in that co-op residents were more strongly committed to staying in their buildings and more involved with their communities than tenants in city-owned buildings or buildings sold through other programs. Co-op residents were more involved in churches, block associations, other neighborhood organizations, and their local government Community Boards. Although it has not been systematically documented, there is also evidence that involvement in the TIL processes teaches residents skills and self-confidence and widens their social networks to include people of diverse education, employment, and access to resources. These expanded skills and contacts often lead to jobs and occupational and educational advancement.

Case Study

The story told by Janice Dozier, a choreographer who moved into 8 West 119th Street in Harlem eight years ago, exemplifies the enormous changes tenants have brought about through the TIL program. "There was prostitution in here, men and women urinating and defecating in the hallways. You had to signal to come up, whistle or shout. And if you took too long, people had to come out and check on you. The police were even afraid to come in here."

"It was crazy people," recalls her son, Vernon. "This was a cracked-out building. Too many monsters."

Half of the tenants in the 20-unit building were senior citizens. Many of the rest were addicted to drugs. The city was the landlord, and the tenants had no control over who the city put in the empty apartments. So they decided to take control of something else: the front door. In 1992, Janice and her partner, Beverly Baillou, began to organize. They created a tenant association and started a security patrol. Five nights a week, between 6:00 and 9:30 p.m., tenants sat in the hallway taking names. The signal was clear; the dealers and addicts could no longer dominate the hallways, and the drug traffic was quickly forced behind closed doors. As the building got cleaned up, the tenant association was accepted into the TIL program. And from that point on, to get a lease in the building, "You had to be on the patrol," says Dozier.

By having control of the rent roll, the association saved up $17,000 in reserves then spent the money on the most needed repairs and on legal fees for evicting six tenants involved with drugs. Today, the hallways are painted, tiled, and clean, with artwork on the walls outside Dozier's apartment. The city has replaced the boiler, the retaining walls, and the stoop and put new kitchens and bathrooms in most of the apartments. They even replaced the roof last summer.

The biggest change of all at 8 West 119th has been security, and it shows on the block outside. "We have cleaned up a little section of this block," says Dozier. "Once people see we respect the block, they begin to respect the block."

The building has become an anchor for the community during the last year and a half. Tenants in this and surrounding buildings have joined with local homeowners and a neighborhood organization, Community Pride, to pull residents together, rebuild a sense of mutual respect and opportunity for the young people in the area, and foster tenant ownership in the other city-owned properties on West 119th Street. Every week, a small group of residents meets in the basement of Number 8. "Those tenants got together and said they would not have it," says Marguerite Gordon, who has owned a house across the street from Dozier's building for 50 years. "It's cleaned up and cleared out. We have all come together now and there's strength."

Both the testimony of tenants and the surveys, two years apart and in two different boroughs, indicate that the TIL program greatly improves housing that was previously devastated. These achievements can be obscured in the day-to-day struggles of maintaining economically marginal housing through a messy democratic process in a context of declining public support for low-income housing. For example, as tax assessments and water and sewer charges increased precipitously in recent years, about one quarter of the co-ops have been in serious arrears in payments to the city and are threatened with foreclosure, a problem that is being worked out for some buildings through caps on water charges; others may well fail. Despite the successes of the TIL program, internal and external threats continue to make these achievements fragile. It is through this juxtaposition of surprising success and constant threat of failure that the lessons learned from the TIL program become clear.

■ Lessons in Program Design and Support

The TIL program succeeded where its 1970s predecessor, the direct sales program, failed for a number of reasons. The development of subsidies, administrative and technical support organizations, and standard procedures for sales, legal transactions, co-op structure, and budgeting made the process possible for more than a handful of especially dedicated tenants. With organizational and financial support and standard procedures more or less in place, the city and the legal and technical assistants who provided program supports learned to deal with many persistent difficulties. Problems came from several different sources, though all of them were conditioned by the tenants' low incomes, as compared to the high cost of providing housing (Saegert, Clark, Glunt, Roane, & Tyler, 1989).

Internal Problems

1. Repair and maintenance. The poor condition of the TIL housing stock burdens these low-income, limited-equity co-ops for the foreseeable future. The last two city administrations have increased the commitment to higher levels of rehabilitation prior to sale, but cutbacks in federal funds jeopardize a major source of capital money (Community Development Block Grants) and ongoing operating subsidies (Section 8 vouchers and certificates). In addition, tenants sometimes encounter structural problems that are beyond their capacity to address. Lack of regular architectural and engineering assessments and the absence of long-range planning often add to the difficulty of keeping the buildings in good shape.

Most of these problems are faced by all forms of housing for low-income residents. However, tenant cooperatives may be particularly prone to postponing expensive long-term investments in favor of keeping rents affordable, avoiding debt, and preserving the basic comfort of the building. Because tenants are not part of the for-profit or not-for-profit real estate industry, it is particularly critical that they have access to training resources and professional assistance, as well as low-interest loan programs.

2. Organizational development and maintenance. Regardless of the income level of tenants, housing co-ops are prone to problems of factionalism, inadequate communication among officers, board members, and shareholders, low levels of participation, and problems of shareholder alienation. TIL co-op shareholders' more limited financial and educational resources and work experiences may increase the difficulty of management. The fiscal and physical marginality of many of the buildings makes the consequences of internal problems more serious. Thus, training and crisis resolution mediation are important technical support services for co-ops. Although it has not been consistently available in the TIL program, training in organizational development and maintenance would also be useful.

3. Management. Whether a building is self-managed or employs a paid manager, its board must be able to understand and make effective management decisions and assure their implementation. All buildings have need of legal, accounting, and financial advice and services. Many boards also need training and assistance in finding and supervising paid managers. Problems with access to individual apartments, transfer of ownership and occupancy, and tenant-created nuisances are often hard to resolve.

4. Resale policy. Many buildings are divided over resale policies, including how much to charge, what tenants to select, and the long-term goals for the building. Unclear, changing, and unenforceable city resale regulations have put shareholders in a difficult position with regard to keeping resale prices low. However, sale prices in most buildings have not posed a strong threat to the low-income stock. This could change if financial pressures on the buildings increase and subsidies dry up.

5. Isolation. During the interim lease phase, TIL buildings receive technical and professional advice, services, and training from the city and from UHAB, the technical assistance organization under contract to the city. After sale, such assistance and services are not routinely available. (The city did contract UHAB to provide co-op support, but the scope of the contract was limited to buildings in serious crises.) This problem has been addressed in recent years by the development of several new types of organizations. UHAB organized a federation of co-ops that provides discounted insurance and legal services plans, a credit union, and bulk purchasing of fuel. With some initial organizing support from UHAB, TIL buildings and co-ops have organized themselves into neighborhood networks and a citywide membership organization to address neighborhood issues like crime and to exchange information, develop solutions to building needs, influence city policy and offer peer support to tenants of other buildings interested in going into TIL (Minott, Rechavi, & Clark, 1995). The Task Force on City-Owned Property has also been a source of information and technical and financial support for the neighborhood networks and the citywide coalition.

Technical Support

The TIL program developed out of an alliance of organizing and advocacy groups, legal aid, and tenants with city government. All of these parties are essential to making the program work, but in the face of different institutional pressures, as well as frequent financial crises, they often clash. It is not clear how these conflicts could be avoided in a time of increasing fiscal austerity in the public sector. However, some lessons can be learned. The existence of a variety of sources for technical assistance allows tenants to seek out what they need when they need it and to avoid getting caught in the institutional crises of any particular nonprofit or government agency. The value of diversity has to be weighed against the need for consistent institutional commitment to the goal of tenant ownership.

In New York, political exigencies, shifting administrative mandates, and financial pressures have resulted in program changes that disrupt tenants' efforts to become self-owned. Thus, government has not been a sufficiently reliably constant source of support. Despite organizational and financial crises, the constancy of UHAB's mandate to support tenant cooperatives has been important. Early in 1996, the city and UHAB decided to break apart the technical assistance and training functions. Technical assistance contracts will be put out for bid to local community organizations. How the training will be supplied is unclear. These changes may well increase the diversity of sources of support, and geographically closer organizations may be more responsive to tenant needs. However, it will also be important to preserve the institutional learning capacity and constant mandate that UHAB has supplied.

Technical assistance to tenant-owned co-ops has some unique features. It should cover the diversity of issues described in the section on internal problems. The nature of the relationship that providers develop with the co-ops is also important. Tenant associations need to go from what is often called a "tenant mentality" to being in charge. They must also learn to establish regular, democratic, and responsible procedures to assure collective decision making and the ability to manage their buildings. The institutional demands of technical assistance providers greatly affect how well these groups are positioned to work with co-ops to achieve the necessary expertise. For example, community organizations that also own and manage housing may not find it in their interest to encourage real tenant independence and control. Governmental agencies and the organizations they contract with may have difficulty acting in the interest of tenants when the interest of the old (city) and new (tenant) owners conflict.

Assistance providers also have to strike a difficult balance between encouraging independence and providing timely and appropriate assistance with the many and often very personal and dramatic crises that tenants experience. The variety of tenant needs can stretch the resources and knowledge of almost any provider.

Probably the most effective form that technical assistance could take would be through a tenant-owned co-op service sector. This would minimize the conflicts of self-interest between community organizations, government, and the tenants, and better match the solutions with co-op needs. The New York experience also suggests that as the number of co-ops grows, leaders and residents can take a larger role in organizing the kinds of assistance they need. At the moment, however, the existing networks and citywide coalition have very limited capacity, are primarily voluntary and relatively unstructured, and lack staffing and resources.

Lessons in the Political Context

Because the TIL program is, by its nature, a difficult sell to many bureaucrats, New York's extreme shortage of affordable housing was a key factor in its creation and sustenance over the years. The private nonprofit and for-profit sectors had limited capacities to absorb the thousands of buildings taken into tax foreclosure by the government, so the city was willing to make a commitment to a more complicated program that focused on building tenant control. The cost savings represented by tenant management while the building was still in city ownership were also an important factor.

Most American cities have no corollary to New York's in rem policies. Instead, municipal governments generally auction off or transfer ownership of properties whose landlords have failed to pay tax debts to a third party. New York is also moving in that direction. Mayor Giuliani has presented legislation in the City Council that will allow the city to bundle the tax liens on certain properties with a high tax debt and sell them to Wall Street investors. For properties that are less attractive, the city will seek to work out debt problems with the landlords, and only in situations where this proves impossible will foreclosure actions proceed. Even then, if the current plan is approved, the buildings will only enter city ownership long enough for all mortgage and tax debts to be wiped clean, then the property will be handed over to a new owner. The city has promised to provide very low-cost rehabilitation loans and grants to the new owners—which in some cases will include tenants—to make the properties attractive.

It is possible that a program very similar to TIL could function without the current in rem mechanism. Another organization, either a membership coalition of co-ops or an independent nonprofit such as UHAB, could retain some oversight and control of the buildings as tenants are trained in management and rehabilitation moves forward.

Tenant ownership and empowerment are concepts that have been appropriated by various political interests in recent years. Former HUD

Secretary Jack Kemp's programs promoting tenant management of public housing come to mind; yet, these programs were drafted from above and depended on the benevolent, permanent overview of government authorities. A true TIL success story, on the other hand, is a complete transfer of ownership and control to low-income people—a result that could only come from a program created and driven by grassroots demands.

There are lessons in the ambiguous politics of the TIL program which are representative of neither classic big government liberalism nor Republican free-market ideology. The program appeals, at least on a rhetorical level, to both Democrats and Republicans in New York City, albeit for different reasons. Progressives like the concept of TIL as part of a larger policy promoting community control of housing, while conservatives seem to be enamored of its "empowerment" aspect and decentralization of power. However, in practice, the fact that the program owes as much of its success to tenants as to government, and the fact that it is not ideologically central to either Democrats or Republicans, has also limited its appeal to either party. Neither party has figured out a way to give the TIL program the appearance of massive change that both have sought to portray as their approach to the in rem housing problem. And as subsidies grow more complicated, the political hurdles become higher.

Rehabilitation subsidies are an essential part of the TIL program. Considering the tenants' very low incomes, it would be pointless to provide them with ownership of a building in degraded condition. Operating subsidies are also important. Congress has ceased approving new Section 8 subsidies, so this resource is disappearing. Alternatives may include an up-front investment by the city to help build a reserve fund that would then help pay the co-op charges of very low-income residents. The proceeds from the sales of vacant apartments on the open market could also be used to subsidize operating costs.

Yet, the primary sales point of any particular government policy in this age is cost effectiveness. The challenge is to draw the bottom line far enough out in time so that the true cost benefits of TIL can be recognized. The impact of TIL on the civic culture and neighborhood stability is clearly beneficial. At a time when creating stable low-income housing is almost impossible, tenant-run cooperatives are a proven strategy for providing decent-quality housing. Unlike government and community groups, co-ops have few external costs; unlike the private sector, co-op "profits" are realized with lower costs and a better standard of housing. If long-term redevelopment is the goal, a tenant ownership model in the TIL vein is a powerful strategy.

The following sources are available for more information:

Urban Homesteading Assistance Board, 40 Prince Street, New York, NY 10012. (212) 226-4119.

Housing Environments Research Group, City University of New York, 33 West 42nd Street, Room 627a, New York, NY 10036. (212) 642-2566.

Tenant Interim Lease Program, Division of Alternative Management Programs, New York City Department of Housing Preservation and Development, 150 William Street, New York, NY 10038. (212) 240-5595.

REFERENCES

Bach, V., & West, S. Y. (1993). *Housing on the block: Disinvestment and abandonment risks in New York City neighborhoods.* Community Service Society of New York.

Blackburn, A. J. (1995, June). *Housing New York City, 1993.* Department of Housing Preservation and Development.

Clark, L. (1995, April). Organizing for ownership. *City Limits.*

Cohn, B. J. (1981, July). *Management needs of small low- and moderate-income housing cooperatives in New York City.* New York: Urban Academy for Management.

Culhane, D., Dejowski, E., Ibanez, J., Needham, E., & Macchia, I. (1994). Public shelter admission rates in Philadelphia and New York City: The implications of turnover for sheltered population counts. *Housing Policy Debate, 5*(2): 107-140.

Dinkins, D. (1993, September 7). Responses to *City Limits* Questionnaire.

Kilbane, T., et al. (1984). *A guide to cooperative ownership.* New York: Urban Homesteading Assistance Board.

Laven, C. (1978, March). *A housing and neighborhood development strategy for city-owned properties.* New York: Task Force on City-Owned Property.

Leavitt, J., & Saegert, S. (1990). *From abandonment to hope, community households in Harlem.* New York: Columbia University Press.

Minott, A., Rechavi, T., & Clark, H. (1995). *Neighborhood network final evaluation: A technical report for the Urban Homesteading Assistance Board.* Center for Human Environments, City University of New York.

New York City Rent Guidelines Board. (1994, December). Tenant income and housing affordability. *Rent-stabilized housing in New York City.* New York: Author.

Peirce, N. R., & Guskind, R. (1993). *Breakthroughs, recreating the American city.* New Brunswick, NJ: Center for Urban Policy Research.

Saegert, S. (1993). Survey of residents of currently and previously city-owned buildings in the Bronx. In M. Cotton (Ed.), *Housing in the Balance.* New York: Task Force on City-Owned Property.

Saegert, S., Clark, H., Glunt, E. K., Roane, W., & Tyler, A. (1989). *Planning for permanence: An evaluation of city-sponsored limited equity cooperative housing in New York City.* New York: Robert F. Wagner, Sr., Institute for Urban Policy, City University of New York.

Sierra, L. F. (1993). The contribution of New York City's task forces and working groups to city-owned housing policies, 1978 to the present. In M. Cotton (Ed.), *Housing in the balance*. New York: Task Force on City-Owned Property.

Stegman, M. A. (1993, June). *Housing and vacancy report, New York City, 1991*. New York: Department of Housing Preservation and Development.

Task Force on City-Owned Property. (1996). Survey of residents of currently and previously city-owned buildings in Brooklyn.

White, A. (1995, May) The big squeeze. *City Limits*.

Part III

Searching for Solutions

HOLLY SKLAR

For decades, while tax money subsidized the development of heavily segregated suburbia and upscale "urban renewal," inner-city neighborhoods experienced disinvestment. They were stripped of jobs and services and redlined by banks, government mortgage programs, and insurance companies in a self-fulfilling prophecy of white flight, devaluation, and decline. Residents of disinvested neighborhoods are widely, wrongly portrayed as incapable and culpable—people with problems but not solutions.

Where community development has taken place, it has generally meant the building of structures and the delivery of services, not organizing for comprehensive community revitalization. Most community development corporations (CDCs) have focused on the product, such as affordable housing, and not on the process of making development resident-driven and holistic. Typically, CDCs are "based in communities," but they are not "community-based."[1] Without community control, physical development can be disempowering—a technical or managerial process that excludes and alienates. Community development in the truest sense is only possible when the community is organized to control development.

In their article on the California Mutual Housing Association (CMHA), Neal Richman and Allan Heskin explore one alternative to the CDC approach. The CMHA is a statewide organization, founded in 1989, that "relies on a political base of lower-income resident groups, organized around housing issues," such as limited equity cooperatives and tenant associations in private and public housing. Resident leadership is a priority. A majority of CMHA's statewide board and two regional councils must be representatives of resident-controlled housing organizations or tenant associations seeking control.

CMHA emphasizes long-term capacity building—for example, training resident associations in property operations and management. It seeks to break down rather than perpetuate the division between technical staff and resident leadership. To build leadership, conduct training, and strengthen the bonds among geographically dispersed members, CMHA sponsors an annual co-op camp and housing institute.

Where the CMHA focuses on housing, the Sandtown-Winchester initiative of Baltimore, discussed by Edward Goetz, is a comprehensive community development project, embracing housing, human services, and economic development. However, unlike the CMHA, Sandtown is not resident-driven. Goetz explains that Sandtown evolved from discussions initiated by Baltimore Mayor Kurt Schmoke with the Enterprise Foundation—a national nonprofit housing organization created by developer James Rouse—and BUILD, a local coalition of church, labor, and community leaders. City and foundation staff have played a leading role in the Sandtown project from the development of transformation goals for the neighborhood to specific action plans and their implementation.

Residents were heavily involved in the Sandtown planning process, and resident participation is encouraged through the creation of block clubs, for example. Residents have also been hired to implement programs, such as prenatal health care and housing rehabilitation. Still, says Goetz, "there is little consensus among actors regarding the degree to which community participation should be the primary focus of attention." Critics argue that the Enterprise Foundation and the city of Baltimore have focused too much on physical development, especially housing, and not enough on community building. "These critics suggest that the process is in danger of losing its community base," says Goetz.

The Dudley Street Neighborhood Initiative (DSNI) of Boston offers an alternative approach that is both comprehensive and resident-controlled. DSNI's experience provides the following lessons.[2] Effective community development begins by recognizing and reinforcing the resources and leadership within the community. Low-income people and communities, like middle- and upper-income people and communities, have a mix of strengths and weaknesses, needs and capacities. Successful community revitalization takes a capacity-building approach rather than an "expert"-driven needs assessment or service-product delivery model. In the words of DSNI member Najwa Abdul-Tawwab, a Boston public school teacher, what's "key about DSNI is the word 'initiative.' It works to help people initiate." At DSNI, resident control is exercised at many levels and mandated in the organizational bylaws. Residents hold the majority of seats on DSNI's elected board that also includes nonprofit agencies, businesses, religious institutions, and others. Like CMHA, DSNI is committed to leadership development and to demystifying the technical tools of development.

Organizing is the renewable energy that powers successful community development. It is through organizing that residents become involved and united, exercise leadership, achieve immediate goals (such as DSNI's first campaign to close illegal trash dumps), and forge a unified vision of the future. Organizing creates the strength and political will needed to translate that long-term vision into reality. DSNI's vision of holistic, sustainable development embraces housing, jobs and commercial development, a healthy environment, cultural diversity, recreation, education, child care, and other human development.

DSNI turned the traditional top-down urban planning process on its head. Instead of struggling to influence a process driven by city government, business, and foundations, Dudley residents and agencies became visionaries, hired their own planning consultants, and created their own bottom-up "urban village" redevelopment plan. DSNI built an unprecedented partnership with the city to implement this resident-driven redevelopment.

Another key to successful community revitalization is development without displacement of current residents. DSNI decided to seek eminent domain authority over 30 acres of vacant land in the most burnt-out part of the neighborhood after realizing that piecemeal redevelopment would be ineffective—and could lead to speculation and displacement. DSNI made history when it became the nation's first neighborhood group to win the right of eminent domain. It established a community land trust to assure long-term community control and housing affordability.

Community revitalization requires significant resources from outside the neighborhood as well as inside. As the Sandtown case study shows, it requires committed public and private partners. Like Sandtown, DSNI was begun with the strong support of a private foundation, the local Riley Foundation. But Riley personnel have never sat on the DSNI board nor otherwise exercised organizational leadership. Riley administrator Newell Flather explains the foundation's role as a "catalytic role" rather than a "controlling one."[3] DSNI has also built an agency collaborative to strengthen agency accountability to residents, minimize competition, maximize cooperation, and strengthen human development resources and policies. Funders can play an important role by encouraging collaboration and respecting the priorities of a resident-driven collaborative process—priorities that may well differ from funders' preconceived ideas.

As Mary Brooks shows in her article on housing trust funds, federal housing assistance has been totally inadequate. The same is true for urban assistance and human services in general. The situation promises to steadily worsen with the continued drive to balance the budget on the backs of low-income Americans. Brooks explains that housing trust funds (e.g., real estate transfer taxes and linkage fees) began in the 1980s as a way to provide a local, relatively permanent source of revenue for affordable housing. She cites property tax-based financing of local schools as one precedent. However, it is important to note that heavy reliance on property tax revenues, which naturally vary greatly between higher-income and lower-income areas, helps perpetuate "savage inequalities" in the nation's school system.[4] Although housing trust funds are an important source of revenue, they cannot replace a sustained state and federal commitment to assuring a "decent home and a suitable living environment for every American family," as promised by the Housing Act of 1949.

As Brooks says, "The single most decisive factor about the ease in getting a housing trust fund passed is the degree of political will behind it." So too, political will is key to generating government action at all levels. Goetz observes, "The irony of [Sandtown] is that the scope of program objectives and the resources that must be harnessed in order to accomplish them make the Sandtown effort difficult to reproduce in many settings."

Yet, if comprehensive revitalization is not reproduced, it will ultimately fail where it has already achieved success. As DSNI's economic development task force observes:

> Neighborhood Economic Development is a complex process which involves nearly all aspects of personal and community life, and is deeply influenced by regional, national, and international forces.
>
> A community group like DSNI must see economic renewal as a long-term process which will require a whole range of efforts and a sustained, focused strategy. No one project, program, or organization will turn the local economy around. . . .
>
> The key to economic renewal for the neighborhood is ultimately neighborhood residents who have greater access to a wide range of opportunity (education, jobs, small businesses) both within and outside of the neighborhood.[5]

The goals of economic renewal challenge national (and international) trends toward downsized jobs and wages and heightened economic inequality and insecurity. Just as conservative ideologues intended, the federal budget deficit—produced by large tax breaks for the wealthy and skyrocketing military spending in the 1980s—has been used as a permanent enforcer of cutbacks in social spending generally and urban reinvestment in particular. In place of a comprehensive approach to community revitalization, the deficit has been used to justify a zero-sum game in which most lose competing for scarcer federal dollars. As a Boston City official observed, "We didn't cause the deficit, but we are paying for it. . . . We have had an increase in the 'robbing Peter to pay Paul' syndrome, in which community health centers are cut to pay for new infant mortality initiatives and public housing programs are cut to increase the number of shelter beds."[6]

True community revitalization depends on collaboration for mutual progress, not a beggar-thy-neighbor competition for increasingly scarce resources. For disinvestment to be reversed and community revitalization to succeed, people must hold government and banks accountable to reinvestment in presently low-income communities and make democracy work from the neighborhood to the national level.

NOTES

1. William Slotnik, director of the Community Training and Assistance Center, Boston, Massachusetts, quoted in Peter Medoff and Holly Sklar, *Streets of Hope: The Fall and Rise of an Urban Neighborhood* (Boston: South End Press, 1994), p. 256.

2. Based on Medoff and Sklar, *Streets of Hope.*

3. Quoted in Arlene Eisen, *A Report on Foundations' Support of Comprehensive Neighborhood-Based Community-Empowerment Initiatives*, sponsored by East Bay Funders, Ford

Foundation, The New York Community Trust, Piton Foundation, Riley Foundation, March 1992, p. 40. Also see Medoff and Sklar, *Streets of Hope*, Chapters 2 and 9.

4. See Jonathan Kozol, *Savage Inequalities* (New York: Crown, 1991).

5. DSNI Economic Development Committee, "Economics With People in Mind: A Summary of DSNI's Approach to Economic Development," 1993.

6. Testimony of Howard Leibowitz, Mayor Ray Flynn's Director of Federal Relations, before the Boston City Council Special Committee on Intergovernmental Relations, February 26, 1991.

8 Sandtown-Winchester, Baltimore: Housing as Community Development

EDWARD G. GOETZ

Abstract

The Sandtown-Winchester neighborhood of Baltimore is undertaking a comprehensive program of neighborhood transformation that features extensive partnerships between the city, community residents, and the Enterprise Foundation—a national nonprofit organization. Called Community Building in Partnership (CBP), the initiative mixes an aggressive strategy of housing rehabilitation and new construction with programs to enhance social services, education, job training, neighborhood beautification, and community building. In this chapter, I analyze the CBP process and describe the multidimensional approach used to transform one of the most devastated neighborhoods in America. The local initiative has made use of public housing rehabilitation funds, the federal Nehemiah Housing program, and local city and state funds to renovate or create over 900 units of housing in the neighborhood with plans for the rehabilitation of 600 more vacant properties.

■

AUTHOR'S NOTE: The author would like to thank the following people who provided information on the Sandtown transformation process: Sara Eilers, Enterprise Foundation; Lenneal Henderson, University of Baltimore; Ella Johnston, Sandtown-Winchester Neighborhood Improvement Association; Edward Landon, Baltimore Housing Authority; Karen Safer, Maryland Department of Housing and Community Development; Sola Seriki, Neighborhood Development Center, Baltimore; Dennis Taylor, Baltimore Department of Housing and Community Development.

The lack of any discernable urban policy at the federal level since the Carter administration has led to a recent proliferation of local community development initiatives funded by foundations, local governments, and the private sector (Eisen, 1992; Jenny, 1993). These programs vary significantly in their scope and methods, though all attempt to reverse the decline of inner-city neighborhoods either through physical renewal, the enhanced delivery of services, or the empowerment of local residents. Among the most comprehensive of these efforts is the Community Building in Partnership (CBP) effort taking place in the Sandtown-Winchester neighborhood of Baltimore. CBP is a comprehensive program of community revitalization, incorporating significant housing rehabilitation and new construction, enhanced health and human service delivery, open space improvements, community building, and job creation. The program was initiated by Mayor Schmoke, a coalition of churches, unions, and community leaders, and the Enterprise Foundation. In this chapter, we examine the CBP process and housing as one element in a comprehensive program of neighborhood transformation.

Baltimore

There is much that is typical of U.S. cities in the recent development history of the City of Baltimore. Dependent upon its harbor and manufacturing to create a foundation of blue-collar jobs, the city grew throughout the first half of the century, until in 1970, over 900,000 people lived within its limits. But similar to other U.S. cities, the metropolitan area grew up around it, fueled by suburban growth and the gradual decentralization of economic activity. The suburban areas surrounding Baltimore were shaped by the exclusionary policies of the Federal Housing Administration (see Jackson, 1985) and became home to middle-class whites, while racial minorities, largely a sizable African American community, were limited in their housing choice to the aging inner-city neighborhoods. The segregationist pattern set in motion by the FHA was preserved through the exclusionary zoning practices of suburban communities that erected barriers to affordable housing, leading in turn to further concentration of lower-income families and African American families in the inner city (McDougall, 1993).

Baltimore has suffered a major loss of manufacturing jobs since 1960. From 1960 to 1986, the number of manufacturing establishments in the city declined by over 60% (Hula, 1990), and between 1970 and 1985, the city lost 40,000 manufacturing jobs or 45% of its manufacturing employment base (Levine, 1987). As jobs left, so did residents; between 1970 and 1985, the city experienced a net loss of 134,000 white households. By 1990, the city's population was down to 736,000, dropping from almost 45% of the metropolitan area's population in 1970 to less than one third in 1990. During the same period, the Black population within Baltimore increased from 46% to almost 60% of the city's total (see Figure 8.1). Among the city's Blacks, the net effect of job and population loss was to leave a more impoverished community. Between 1969 and 1989, the black median family income as a percentage of the metropolitan area's income fell from 69% to 55%.

In the 1970s and 1980s, the City of Baltimore undertook an aggressive campaign of downtown-centered economic redevelopment in an attempt to reverse its decline. Under the leadership of an organization of business executives called the Greater Baltimore Committee (GBC) the city initiated the Charles Center project, 33 acres of office space, retail, and commercial developments in the central business district (CBD), and the Inner Harbor project that included more commercial and retail space along the city's waterfront. The pace of redevelopment increased with the election of Donald Schafer as mayor in 1971 (Levine, 1987). Under Schafer, the city focused on downtown redevelopment by accelerating the Inner Harbor project and adding a number of publicly financed developments scattered throughout the CBD. These projects were pursued despite an overall retrenchment in city government during this time. Though the city's overall budget was reduced by 20% during the 1980s, economic development expenditures increased by over 400% (Levine, 1987). So active was the city's business leadership in downtown renewal, and so numerous were the quasi-public and private associations formed to facilitate downtown development, that they were called Baltimore's "shadow government" (Levine, 1987; Berkowitz, 1984). All this "pump-priming" of the downtown market by the city paid off when James Rouse completed his Harbor Place development, a historically renovated marketplace on the waterfront, in 1980. This, in turn, spurred another round of private sector investment along the harbor and in the central business district (Levine, 1987). The success of these projects was so dramatic that Baltimore became a national model for successful inner-city revitalization.

The economic turnaround of the downtown area did not, however, extend beyond the central business district. In the neighborhoods of Baltimore, a different scenario was playing out. The white middle class continued to leave the city, the black middle class fled inner-city neigh-

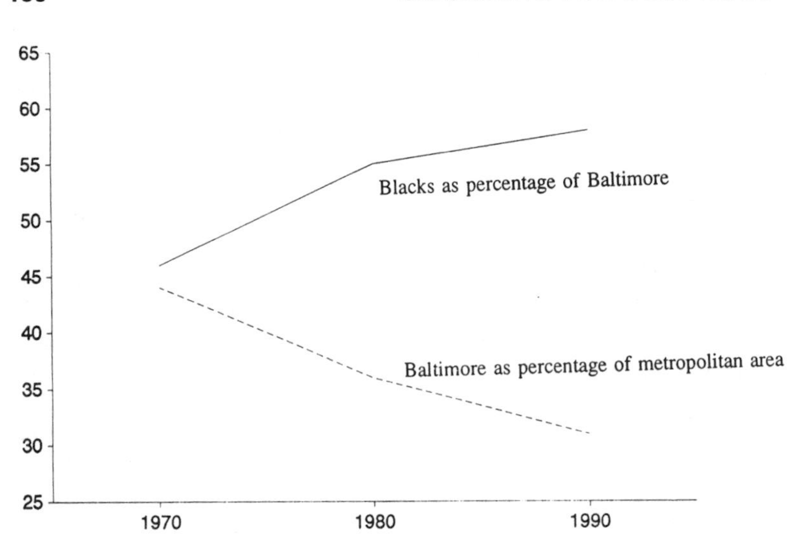

Figure 8.1. Population Changes in Baltimore Metropolitan Area, 1970-1990

borhoods, and once stable inner-city neighborhoods sank into landscapes of neglect. Three of every four black neighborhoods in the city experienced an increase in the rate of poverty between 1980 and 1990 (Levine, 1987). Lending institutions continued their pattern of disinvestment and the housing stock deteriorated, while costs increased. Despite the growing evidence of neighborhood disinvestment, those running Baltimore's economic development policy continued to focus on downtown projects, the benefits of which went primarily to developers and suburban professionals (Levine, 1987). The jobs created downtown required either advanced training unavailable to the city's minority population or were low-paying, low-status service sector jobs incapable of supporting families. By 1986, in a report titled *Baltimore 2000,* Szanton (1986) described Baltimore as a city with two distinct realities, the vibrancy of downtown and suburban areas that offered employment and recreational opportunities to a primarily white population, and the neglect, poverty, and physical deterioration that characterized the inner-city black neighborhoods (see also Hula, 1990). This report helped to politicize the distribution of benefits from the strategy of downtown redevelopment pursued by the city. When Mayor Schaefer vacated his seat to run for governor, a black city councilmember, Kurt Schmoke, ran for mayor and was elected partly on his promise to deliver more to the city's neighborhoods. Shortly after taking over, Schmoke began to make good on his promises by focusing attention on what was perhaps the most deteriorated neighborhood in the city, Sandtown-Winchester.

Sandtown-Winchester Neighborhood

During the 1940s, Sandtown-Winchester was a solid, working-class neighborhood, the commercial center and heart of Baltimore's African American community. The boyhood home of Supreme Court Justice Thurgood Marshall, Sandtown was where the black community shopped at black-owned stores along the Pennsylvania and North Avenue commercial strip. The neighborhood had an exuberant night life highlighted by the jazz clubs along Pennsylvania Avenue. But the disappearance of blue-collar jobs, the closing of local factories, and the subsequent flight of the black middle class out of Sandtown pushed the neighborhood into rapid decline. By the end of the 1980s, Sandtown was, according to a local pastor, "like the third world; not a night goes by that I don't hear the police helicopter overhead" (Gunts, 1991). When James Rouse was looking for a neighborhood in which to start a model community revitalization program, a neighborhood "with real devastation, where people were really living in the worst possible conditions . . . a bottom neighborhood," he chose Sandtown (Gugliotta, 1993).

Sandtown is a 72-square-block area on the west side of Baltimore only 15 blocks from downtown (see Figure 8.2). The population of the neighborhood is almost completely black and has declined in number since 1960. Though home to over 30,000 people during its heyday, the current population is just over 10,000 (Miripol, 1994). By the late 1980s, the neighborhood was choked by problems of crime and physical neglect, characterized by massive disinvestment, both public and private, absentee landlords, abandonment and vacant lots, poor housing conditions, open-air drug and prostitution markets, and pervasive poverty. The neighborhood's infant mortality rate was twice that of the city, the incidence of HIV infection was three times that of the city, and 90% of children born in the neighborhood were born to single mothers (Pugh, 1993). One half of the households had incomes below the poverty level, and 45% of the neighborhood's households received public assistance. Half of the working-age adults were unemployed or underemployed (Gugliotta, 1993).

Figure 8.2. Sandtown-Winchester Neighborhood

The housing stock in the neighborhood is old and deteriorated. In 1990, 75% of the housing units had been built prior to 1949. The area was littered with vacant, boarded-up properties and vacant lots overgrown with weeds. According to estimates, there were 600 vacant properties in the neighborhood and 3,000 structures in need of rehabilitation (McDougall, 1993).

Of the 4600 housing units in Sandtown, 70% are single-family homes, though only 20% of those are owner-occupied. Rents, though inexpensive by city standards, are still unaffordable for a large portion of Sandtown residents. Home buying is also beyond the means of most Sandtown residents, despite the low cost of row houses in the neighborhood (Olesker, 1993). The neighborhood is the site of Gilmor Homes, a 571-unit low-rise public housing development; Harvey Johnson Towers, a 120-unit high-rise elderly housing project financed by the state; the Frederick Douglass Apartments, 100 rental units in a converted high school financed by a federal Housing Development Action Grant (HoDAG) in 1987; and other smaller subsidized housing projects.

Community Building in Partnership

Shortly after his election in 1987, Mayor Kurt Schmoke initiated discussions with the Enterprise Foundation, a national nonprofit housing organization, and BUILD, a local coalition of church, labor, and community leaders, that led to the city submitting a proposal to HUD for assistance under the Nehemiah Housing program for the development of 300 new homes in the West Baltimore neighborhoods of Sandtown and Penn North. This initial collaboration led to the more comprehensive CBP process that went beyond housing to target education, crime, employment, social service, and empowerment issues in the Sandtown-Winchester neighborhood.

BUILD

A united church-labor-community coalition called Baltimoreans United In Leadership Development (BUILD) is a major player in the creation of the Sandtown project. BUILD is an Alinsky model coalition of 45 churches, three local unions, and residents of public housing. The Coalition that emerged from the Baltimore Interfaith Ministerial Alliance in the early 1980s pursued a high-profile strategy of confrontation during Mayor Schaefer's push for large-scale downtown revitalization during the early 1980s. It soon became the premier voice of the poor in the city. The organization has enjoyed a closer relationship with City Hall since the election of Kurt Schmoke in 1987 (McDougall, 1993).

Though BUILD was the leading community-based organization in the formation of CBP, community interests are also represented by the

Sandtown-Winchester Improvement Association (SWIA). SWIA was formed in 1977 and has served as the citizen participation organization for the neighborhood in previous renewal efforts. SWIA spun off the Sandtown-Winchester Community Development Corporation in 1985 to be the development arm of the community.

City of Baltimore

The large-scale economic development strategy pursued by the City of Baltimore began to change during the 1980s, as federal subsidies were cut. Between 1980 and 1990, federal government funds as a percentage of the city's total budget shrank by 50%, reflecting the loss of development subsidies and necessitating a reorientation in development strategies. The cutback in federal funds also restricted funding to community organizations, leading to conflict between citizen organizations over development (McDougall, 1993). The election of Schmoke signaled a change in concern at City Hall that favored neighborhood-defined development objectives as part of a community-based revitalization strategy. In 1988, for example, the Mayor created the Community Development Finance Corporation (CDFC), a community development bank formed to make housing rehab loans in the city. Later in that year, the Mayor visited the Israeli city of Kiryat Gat and returned hoping to initiate a similar program of self-maintained, self-sufficient community development in Baltimore.

Enterprise Foundation

The final large partner in the Sandtown-Winchester redevelopment story is the Enterprise Foundation (EF), a national nonprofit organization created in 1981 by James Rouse. Rouse is the nationally renowned developer responsible for high-profile downtown commercial developments, such as Fanueil Hall in Boston and Harbor Place in Baltimore. His success in transforming old and tired downtowns into vibrant commercial districts led him to experiment with the idea of similarly revitalizing older inner-city neighborhoods. The Enterprise Foundation is a national intermediary, channeling investment and foundation funds to groups working to preserve and develop affordable housing across the country. By 1990, EF had helped 130 neighborhood groups in 60 cities develop over 15,000 houses.

In 1988, Schmoke, EF, and BUILD formed a three-way partnership to present a proposal to the federal government for funding housing revitalization through the Nehemiah Housing program. The Nehemiah program requires a large-scale housing renewal effort that provides homeownership for low-income households but also creates major neighborhood revitalization through clearance and new construction. The city's Housing and Community Development Department had already acquired enough land in the Sandtown-Winchester neighborhood to make such a project feasible (McDougall, 1993). The Sandtown Nehemiah proposal was funded by HUD in the amount of $4.2 million in 1989; the largest Nehemiah grant made by the federal government that year. EF created a subsidiary organization, the Enterprise Nehemiah Development Corporation, to develop the housing.

The creation of the Sandtown CBP process resulted from a confluence of emerging objectives between BUILD, the Mayor, and EF. BUILD saw the Schmoke administration as more amenable to neighborhood concerns, and they wished to correct years of public neglect and private disinvestment in Sandtown. Schmoke himself was interested in developing a neighborhood-oriented development strategy and creating a self-sustained community development effort in the city and channeling more public resources to neighborhoods. At the same time, the Enterprise Foundation was having internal discussions about the possibility of focusing a concentrated amount of assistance in a single community to develop a model of comprehensive neighborhood revitalization. Using the Nehemiah project as a springboard, the three parties began talks about a more comprehensive community development approach that would go beyond housing rehabilitation and would incorporate significant community participation and leadership.

■ The Process

Phase I

Schmoke initiated Community Building in Partnership by creating a task force in January of 1990 that included representatives from the city, BUILD, Enterprise, the Baltimore Urban League, and the Sandtown-Winchester Improvement Association (SWIA). The first phase of the CBP process was the development of community goals. In May of 1990, the task force began their work with a public meeting attended by 400 people. Using the input from the mass meeting, the task force convened eight work groups in October, 1990, to establish goals within more specific topical areas. The eight work groups focused on the following:

1. Substance abuse
2. Physical development
3. Health care
4. Education
5. Employment and community economic development
6. Family development
7. Public safety
8. Community pride and spirit

Each of the work groups included representatives from the Enterprise Foundation, city staff, and community members. The product of their work was a series of transformational goals for the neighborhood issued in April, 1992.

Phase II

Phase II was begun when clusters were formed to translate the transformational goals into specific action plans. Four program design clusters were created that consisted of residents, city staff, Enterprise Foundation staff, and practitioners in the field. Each was led by a professional facilitator. The clusters focused on human services, education, physical and economic development, and community building. Each cluster prepared a set of specific programs, action steps, and resources needed to implement the vision of the first phase of planning. The final plan, which was made public in March of 1993, called for a range of programmatic activity across all cluster areas.

Program to Transform Sandtown-Winchester[1]

The final plan for the transformation of Sandtown-Winchester incorporated a wide range of strategies. In the area of *physical and economic development,* the plan called for the following:

- creation of a nonprofit neighborhood development center to assist developers and homeowners to assemble financing, address property management needs, and oversee development of open space;
- renovation of a neighborhood marketplace;
- revitalization of the Pennsylvania Avenue commercial strip;
- establishment of a Neighborhood Employment and Training Center;
- development of a community improvement jobs training program; and
- creation of an industry-linked training program with regional employers.

In the *health and human services* cluster, the plan called for the following:

- creation of an intensive professional development program for public school employees;
- connecting public schools to pre-school, tutoring, mentoring, and other community-based support services;
- use of schools as child development centers and sites for community education;
- linkage of outreach and delivery of human services through family support teams;
- creation of a centralized location for service delivery and integration;
- creation of a consortium of service providers for coordination and integration of services;
- expansion of services for youth in the neighborhood;
- provision of a full range of health services to residents;
- creation of a community-based health care system focused on prevention and primary care;
- creation of a range of substance abuse treatment services in the neighborhood; and
- provision of substance abuse prevention programs in the schools.

The *public safety* goals included the following:

- increase in resident involvement and responsibility in crime prevention programs; and
- implementation of community-oriented policing.

Finally, in the area of *community building,* the plan called for the following:

- creation of a coordinating council of community organizations;
- establishment of a volunteer clearinghouse to train and place volunteers; and
- development of recreational, cultural, and historical programs to nurture community pride and spirit.

Phase III

The third phase of the CBP process is implementation and evaluation. Community Building in Partnership, Inc., was formed in 1993 as a nonprofit organization responsible for the coordination of implementation activities. Much of the actual program implementation will be done by existing service agencies.

The process of rebuilding Sandtown-Winchester has not been as linear as the careful, graduated method of goal identification, action plan development, and implementation planning described above would suggest. Almost from the beginning of the planning process, the Enterprise Foundation staff emphasized the need to make immediate improvements in the built environment of the neighborhood to generate enthusiasm around the planning process and to attract foundation and public funding. In fact, CBP, Inc., officials count the Nehemiah project as the first accomplishment in the CBP process, though it actually predated the planning process. Other housing efforts were also begun by Enterprise and other nonprofits prior to the completion of the neighborhood plan. By November, 1993, only seven months after the completion of the action plan, over $60 million in government and foundation funds had already reached Sandtown (Bock, 1993b).

Program Areas

Housing

Housing programs have dominated the physical rehabilitation of Sandtown-Winchester. As described earlier, the neighborhood housing stock was in extreme disrepair. The neighborhood was glutted with abandoned buildings and vacant lots and burdened by problems of arson and absentee ownership. Thus, Enterprise emphasized housing rehabilitation (its area of expertise) from the beginning of the CBP process, and it is this area in which the greatest progress has been made in the neighborhood.

Nehemiah. The Sandtown Nehemiah project consisted of 227 townhouses (210 new construction and 17 rehabilitated) built on the site of one of the largest blighted properties in the neighborhood, an old abandoned bakery. The townhouses are two-story, three-bedroom units that are sold for as little as $256 per month. The City of Baltimore contributed $11.1 million in land, site clearance, and infrastructure to the project. BUILD provided $2.2 million in construction loans to the developers, while the HUD Nehemiah program provided $4.2 million in subsidies to the homebuyers. The subsidies came in the form of interest-free second and third mortgages of up to $25,000. If a purchaser remains in the house for 10 years, these subsidies are forgiven; they are due and payable upon sale of the property within 10 years.

This initial attempt at affordable housing, however, did not seem to be targeted to incomes low enough to affect the average Sandtown resident. The average annual income of purchasers was $17,000, well above the median household income for the neighborhood of $12,465 (Bock, 1993b). Almost 90% of the buyers were single-parent households, though the average household size was just 1.5 persons. Half of the purchasers had previously lived in Sandtown or nearby neighborhoods. In the Nehemiah project, 73 more units were built in the Penn North neighborhood directly north of Sandtown. The project is supported by homeownership counseling, budget management, and home maintenance workshops provided by the Enterprise Nehemiah Development Corporation. The units are fully occupied with an active homeowner's association.

Habitat for Humanity. The Sandtown-Winchester Habitat for Humanity chapter was formed in 1988 by volunteers who also formed the New Song Church in Sandtown. In 1992, Sandtown Habitat officials announced a goal of building 100 homes in Sandtown in a five-year period. To help publicize the event and raise funds for materials, Jimmy Carter visited the neighborhood in March, 1992. In June of that year, Habitat kicked off its Sandtown campaign by conducting a "blitz build," in which hundreds of volunteers built ten homes in one week. The blitz build was repeated in July of 1993 and again in July of 1994. Habitat provides the homes to low-income families through an interest-free mortgage that covers the cost of materials. In 1993, Habitat homes sold for an average of $30,000 (*Baltimore Evening Sun,* 1993a). New Habitat homeowners had an average annual income of $10,000 in 1992 (Bock, 1992). As of the summer of 1994, Habitat was one third of its way to the goal of 100 units.

Gilmor Homes. Gilmor Homes is a 571-unit, low-rise, family public housing project that takes up four square blocks in the northern half of Sandtown. In 1987, the Baltimore Housing Authority received a $13.3 million grant from HUD through its Comprehensive Improvement Assistance Program (CIAP) program designed to modernize older public housing projects. The grant funded total renovation of the units, including the conversion of 16 one-bedroom units into two-bedroom apartments. The cost of renovation was $23,594 per unit. Currently, the Baltimore Housing Authority and CBP are funding tenant organizing efforts at Gilmor Homes.

Vacant Housing Initiative. At the unveiling of the Sandtown action plan in March 1993, Mayor Schmoke went beyond the provisions of the plan to promise that within one year all 600 vacant or abandoned properties in the neighborhood would be rehabilitated (Fletcher, 1993). Though this declaration proved impossibly optimistic, it did launch a large-scale effort by CBP, Inc., to deal with the neighborhood's vacant properties. A new coordinating agency, the Neighborhood Development Center (NDC), was organized to oversee the purchase and renovation of the vacant properties. NDC is being supported by $1.5 million in grants over a three-year period from the National Community Development Initiative (NCDI), a consortium of national foundations that have pooled contributions to provide grants and loans in support of nonprofit housing development. The work plan calls for 140 small, row-house properties located on alley streets to be demolished and made into public space for gardens or parks. The Initiative also calls for 300 structures to be renovated for rental housing and 300 units to be rehabilitated for resale to low-income buyers. One hundred units of new construction are also planned.

Properties are acquired in three ways: through the city's use of eminent domain after condemnation (ownership is transferred to the city's real estate division of the Housing and Community Development Agency), through direct acquisition from the owner, or through abandoned properties that have gone into receivership. In September, 1993, the city's Housing and Community Development (HCD) Department provided a $627,500 grant to the Community Law Center to provide the legal assistance to put properties into receivership (Simmons, 1993). All properties that are acquired are put into a central land bank accounting system, and developers are assigned. Eight different for-profit and nonprofit developers will do the construction and rehabilitation work.

All properties in the program are row houses, typically three-bedroom units. The resale units will be sold for $90,000 but will cost buyers only $39,000. The difference is made up by "silent" second and third mortgages that are forgiven if the buyer remains in the house for 10 years. The second mortgages remain intact if there is resale to an income qualified buyer within the 10-year period; otherwise, the second and third mortgages are due at resale. As of this writing, no new construction has occurred yet, nor have any of the homeownership units been completed. Eighty units have been renovated and rented thus far. Income limits for renters are at 60% of the median adjusted for household size. The demolition of the 140 alley properties was estimated by HCD to cost $615,900 (Daemmrich & Simmons, 1993).

NDC officials estimate that the entire vacant housing initiative will cost $63.5 million, including $2 million from HUD under the Housing Opportunities for People Everywhere (HOPE) program, $20 million from the State of Maryland, $4 million in city bond offerings, and the balance from CDBG. In March, 1994, the city acquired an additional $8 million commitment from Nation's Bank, $2 million for purchase of the homes, and $6 million for a revolving loan fund for financing purchase by homeowners.

Other Housing Developments. In addition to the major housing initiatives described above, two smaller projects have been undertaken since the beginning of the CBP process. The City Homes program of the Enterprise Foundation has funded the rehabilitation of 30 units in Sandtown. The program provides lead abatement for rental properties purchased by Enterprise that require only moderate rehabilitation ($20,000 to $30,000 per home). Enterprise then rents the units for less than $275 per month. The program is funded by the Enterprise Foundation, NCDI, and the State of Maryland Community Development Agency.

Sandtown Manor, 11 units of newly constructed low-income rental housing, was completed in 1992. Units are rented to households at or below 30% of the area's median income. The project was completed by the Sandtown-Winchester Community Development Corporation (SWCDC) with $740,000 in funds from the State of Maryland, the city's CDBG program, and the City of Baltimore's Community Development Finance Corporation.

Human Services

The CBP plan for Sandtown-Winchester combines aggressive housing rehabilitation with a wide range of human services and public space development. In the areas of health, education, and job training, CBP, Inc., has initiated and expanded a range of services for Sandtown residents.

In 1990, the Baltimore Health Department, the Mayor's office, and the NIH began the Baltimore Project, a program of prenatal health care designed to reduce infant mortality in the city. The program began with $650,000 in funding and 17 employees. After two years, the Baltimore Project received a federal grant under the Healthy Start program and began to hire local residents to provide the outreach for the program. In 1993, another $5 million grant from the federal government allowed expansion of the program beyond Sandtown into other neighborhoods in Baltimore. The program is now located in the old Douglass High School building, providing preventive health care for 600 pregnant women each year and employing 70 people.

Five million dollars of public and private funds are also sponsoring the creation of a Health Clinic in Sandtown, scheduled for opening in 1995 (Bor, 1994). At New Song Community Church in the heart of Sandtown, the church provides a learning center for children, a health clinic, and jobs program. The Mount Street Support Center, located in the southern half of the neighborhood, provides a one-stop source for government services from job training to youth recreation. The public schools have also made a commitment to providing programs for children at the three elementary schools in Sandtown (Bock, 1993b). Parent Centers have been created in the three elementary schools to enhance parent involvement. Other programs funded by the Baltimore City Public Schools, the State of Maryland, and the U.S. Department of Education are providing before- and after-school programs for Sandtown schools.

Public Spaces

The community has initiated anti-crime programs and neighborhood beautification programs. In February of 1994, residents and police swept through the neighborhood cleaning alleyways, boarding up buildings in an attempt to leave fewer places of business for drug users and dealers (Jacobson, 1994). In the summer of 1993, the neighborhood and police sponsored a weapons exchange, offering $25 for each gun brought in. The anti-crime movement in Sandtown was kicked off by a "Hands Across

Sandtown" event, in which residents joined together to physically link themselves from one end of the neighborhood to the other (Maushard, 1993).

CBP, Inc., has organized community gardens in abandoned lots, and the city has sponsored tree planting and beautification projects throughout the neighborhood. Four Sandtown parks and recreation centers have been renovated as well, with over $500,000 in funding coming from the Baltimore City Department of Parks and Recreation and the National Park Service.

Community Building

City and Enterprise Foundation officials, as well as community residents, each argue that the success of Sandtown's revitalization ultimately lies in the extent to which community residents mobilize to direct the transformation process. Unlike other well-known revitalization efforts, such as the Dudley Street Neighborhood Initiative (DSNI) in Boston (see Medoff & Sklar, 1994), the Sandtown effort has always had a large element of city and foundation direction. Yet, all actors acknowledge the importance of community building to providing lasting foundation for changes being made in the neighborhood.

Though the neighborhood is in an advanced stage of decline, its history provides it with a foundation for building community. The reunion of old Sandtown residents that takes place each year reflects the commitment that at least a portion of current and former residents have to Sandtown as a community. In 1991, current and former residents of Gilmor Homes, the public housing project located in the neighborhood, also began a tradition of resident reunions. By 1994, over 300 current and former residents of the project attended (Heard, 1994).

CBP has attempted to create a deeper level of community building both through the planning process and during the implementation stages. During the planning process, each work group was heavily represented by community residents; indeed, the whole process was meant to be community driven. In December, 1993, Enterprise Foundation announced that it was receiving a commitment from VISTA for 17 volunteers to work with CBP. The volunteers were recruited from current residents of the neighborhood. These volunteers almost doubled the staff size of CBP, Inc., and helped with job and youth programs and parks and beautification (Bock, 1993a). In addition, community residents have been hired wherever possible to implement programs sponsored under CBP, such as in the Healthy Start prenatal health program.

Community building is also taking place through the creation of block clubs. By March, 1994, half of the 280 blocks in the neighborhood had volunteer block captains who coordinate or carry out block watches and organize neighbors (Hilson, 1994). Organizations like SWIA, BUILD, the Gilmor Homes Tenants Association, and the Nehemiah Homeowners' Association have received funding or provided in-kind support for organizing residents and linking them to the CBP process, as well as to services and resources in the neighborhood.

Various events, such as Volunteer Appreciation Day, the Community Arts Festival, and a Pumpkin Patch Celebration, have also been sponsored by CBP in an effort to unite residents and build community pride.

Evaluation

To a great extent, it is still too early to provide a cogent evaluation of the Sandtown Community Building in Partnership effort. The implementation of programs and projects is quite recent, and it is too early to tell if the interventions will have a significant impact on the neighborhood's future.

In 1994, a team headed by the Conservation Company of New York with the University of Baltimore as the local partner were funded by the Enterprise Foundation to monitor the progress of the Sandtown program. The evaluation focuses on three areas: (a) how much change occurs relative to baseline indicators of community health and well-being, (b) the progress of the program relative to the stated objectives of the plan, and (c) the degree of community-building achieved during the process.

The evaluation is funded through a one-year grant from the Ford Foundation. However, because of delays in choosing a local partner, the evaluation did not begin until May of 1994, and the contract expired at the end of that year. This short time frame did not allow much more than a series of observations regarding the progress of the program. Evaluators were hoping to secure continued funding for the evaluation.

Though neighborhood level statistics on infant mortality, school dropout rates, employment, and other concerns can be collected, attributing a causal connection between change in these indicators and the Sandtown revitalization program will be difficult. Judging progress of the program relative to the plan objectives is straightforward. The final area of evaluation, the degree of community building, is difficult to assess, because there is little agreement about what it means or how it is best measured. The evaluators are not incorporating any measures of cost effectiveness into their study.

The removal of the blighted bakery and the development of the Nehemiah homes has significantly improved the housing stock in the neighborhood. The momentum created by the Nehemiah and Gilmor Homes projects has been continued with the vacant housing initiative. These programs promise to provide immediate improvements in the physical environment of the neighborhood and are the most tangible evidence of improvement in the neighborhood so far.

There are noticeable changes in other areas as well. The police report that many of the open-air drug markets have moved out of the center of the neighborhood to its fringes (Hall, 1994). The overall crime rate in Sandtown has been reduced, though there are still high-profile violent crimes occurring in the neighborhood (Bock, 1993b). Despite the improved crime statistics, there remains tension between community residents and the police. The death of a black Sandtown resident after his arrest in July, 1994, sparked an internal investigation into police procedures and heightened distrust and tension between black community residents and police.

Physical Development Versus Community Building

Some residents and observers complain about the development focus of the Enterprise Foundation and suggest that physical redevelopment has run ahead of community building (Miripol, 1994). There is little consensus among actors regarding the degree to which community participation should be the primary focus of attention. Critics argue that the Enterprise Foundation has focused too much on physical development (especially housing), because that is what they know how to do best and because of the need for quantifiable accomplishments to attract additional program funding from foundations and the public sector (McDougall, 1993). These critics suggest that the process is in danger of losing its community base and that more attention needs to be paid to community building and less to the building of housing. For its part, the Enterprise Foundation continues to stress the need for large-scale improvement of the housing stock and the physical environment in Sandtown. Pat Costigan, who heads the Enterprise Foundation's Sandtown effort, has argued that the pace of development has to be quick and that community-driven processes like

Habitat for Humanity, though laudable, are simply not on a scale required by the community (Bock, 1993b). In addition, there are those in the community anxious to see some tangible results. As one Sandtown pastor said, "We've been meeting to death. Gardens are nice, the (community) newspaper is nice, but in three years the folk just expect more." This is an issue that has not yet been resolved in Sandtown, though at this point, it does not threaten the continued progress of the transformation process in the neighborhood.

■ Implications

James Rouse calls the Sandtown project "the most important single development in the U.S. today, to change the life [sic] of poor people and the heart of the American city" (Bock, 1994). This program, from the standpoint of the Enterprise Foundation, has been self-consciously presented as a model for export to and use in other cities across the nation. Already, the Enterprise Foundation has begun a similar effort in the Overtown neighborhood in Miami. Enterprise's efforts in Sandtown and Overtown join a new generation of locally initiated community revitalization programs in cities across the country (Jenny, 1993; Eisen, 1992). The Sandtown-Winchester project stands out (along with the Atlanta Project and the Dudley Street Neighborhood Initiative) among these local initiatives for the scope of revitalization envisioned (Jenny, 1993). The Sandtown project aims at nothing less than total neighborhood transformation and hopes to redirect resources from the federal, state, and local levels in both the public and private sector, encompassing housing, human services, employment, education, health, and community pride. The scope of these objectives and the resources that must be harnessed to accomplish them make the Sandtown effort difficult to reproduce in many settings.

The vast amount of private and public funding will be difficult to generate in other locales. The Sandtown project has been able to attract funding, like the Nehemiah grant, that is unlikely to be a major factor in other projects nationally simply because of low or no funding at the federal level. The $13.3 million CIAP grant received by the Baltimore Housing Authority also was an anomaly, an extraordinary grant that is unlikely to be repeated in most places. Even some of the private foundation funding is extraordinary; the $1.5 million in funds from NCDI is from a program that is available only to a limited number of cities nationwide.

The State of Maryland has channeled millions of dollars into Sandtown for a variety of purposes, from education to housing and youth programming to human services. Most cities would find it difficult to garner that level of support from their state governments. The city's successful proposal to have Sandtown designated as an "empowerment zone" highlights the difficulty in funding such comprehensive community revitalization efforts; the federal government funded only six such zones in the entire country.

The very scale of the initiative in Sandtown has also created some political repercussions for the Schmoke administration locally. Leaders in other declining neighborhoods in the city are openly resentful of the money and attention given to Sandtown and have called for greater dispersion of funds so that more neighborhoods might benefit (*Baltimore Evening Sun,* 1993b). This pressure to spread funding around a number of neighborhoods has been well documented in other community development programs (e.g., Dommel, Bach, Liebschutz, Rabinowitz, & Associates, 1980) and will be a threat to the effective targeting of a large-scale intervention, such as the Sandtown CBP process.

At the same time, the scale of the Sandtown effort and its integrative approach is increasingly necessary to reverse advanced decay in inner-city neighborhoods, especially in an era of federal neglect. Sandtown does present itself as a successful model of intergovernmental cooperation and public/private partnership in community development.

The degree of community-based input into the planning process is an important element of the Sandtown project. Though not as community-driven as the Dudley Street Neighborhood Initiative in Boston, the Sandtown transformation process has incorporated a great deal of community involvement, at least at the planning stage. Community residents and representatives of community-based organizations, such as BUILD and SWIA, were involved in all stages of the planning process. The work groups were formed as a result of an initial "visioning" process that took place at an open community meeting attended by over 400 residents (McDougall, 1993). This type of community involvement is probably more important in comprehensive neighborhood transformation efforts, such as Sandtown, than it is for more specific or single-issue projects that focus on a single set of conditions. The continued success of the Sandtown effort rests on the ability to balance concrete physical renewal (as exhibited in their housing achievements to date) with the community building necessary to facilitate resident empowerment and true neighborhood transformation.

NOTE

1. This section is taken from Community Building in Partnership (1993).

REFERENCES

Baltimore Evening Sun. (1993a, July 22) Miracles in West Baltimore, p. 22A.

Baltimore Evening Sun. (1993b, September 27). Life beyond Sandtown-Winchester, p. 6A.

Berkowitz, B. L. (1984). Economic development really works: Baltimore, Maryland. In R. D. Bingham & J. P. Blair (Eds.), *Urban economic development.* Beverly Hills, CA: Sage.

Bock, J. (1992, June 15). Habitat for humanity helps families make a house a home. *Baltimore Evening Sun,* p. 1D.

Bock, J. (1993a, December 3). VISTA to aid in revival of Sandtown. *Baltimore Morning Sun,* p. 4B.

Bock, J. (1993b, November 23). Hard part still ahead for Urban Lab Sandtown: Hope on the horizon. *Baltimore Morning Sun,* p. 1A.

Bock, J. (1994, January 28). 15 Sandtown VISTA recruits get Washington pep talk. *Baltimore Morning Sun,* p. 3B.

Bor, J. (1994, January 21). Health outreach set for liberty. *Baltimore Morning Sun,* p. 1B.

Community Building in Partnership. (1993, August). *Progress report.* Baltimore, MD: Author.

Daemmrich, J., and Simmons, M. (1993, September 24). Vacant rowhouses to be demolished. *Baltimore Morning Sun,* p. 3B.

Dommel, P. R., Bach, V. E., Liebschutz, S. F., Rubinowitz, L. S., & Associates. (1980). *Targeting community development.* U.S. Department of Housing and Urban Development. Washington, DC: Government Printing Office.

Eisen, A. (1992). *Comprehensive neighborhood-based community-empowerment initiatives.* New York Community Trust.

Fletcher, M. (1993, March 7). Mayor pledges changes in Sandtown. *Baltimore Morning Sun,* p. 6B.

Gugliotta, G. (1993, January 4). Rebuilding a community from the bottom up. *Washington Post,* p. A1.

Gunts, E. (1991, July 21). Home sweet first home: How the Enterprise Foundation is helping the working poor buy homes and save their neighborhoods in the process. *Sunday Sun Magazine,* p. 9.

Hall, W. A. (1994, September 27). Sandtown gleam typifies successful fight of blight. *Baltimore Evening Sun,* p. 2A.

Heard, E. (1994, July 17). "Old stomping ground" draws residents. *Baltimore Morning Sun,* p. 2B.

Hilson, R., Jr. (1994, March 13). Effort to revive Sandtown relies on curbing crime. *Baltimore Morning Sun,* p. 3B.

Hula, R. C. (1990). The two Baltimores. In D. Judd & M. Parkinson (Eds.), *Leadership and urban regeneration: Cities in North America and Europe.* Newbury Park, CA: Sage.

Jackson, K. T. (1985). *The crabgrass frontier: The suburbanization of the United States.* New York: Oxford University Press.

Jacobson, J. (1994, February 5). Sandtown-Winchester takes broom to dirt, drugs. *Baltimore Morning Sun,* p. 1B.

Jenny, P. (1993). *Community building initiatives: A scan of comprehensive neighborhood revitalization programs.* New York Community Trust.

Levine, M. V. (1987). Downtown redevelopment as an urban growth strategy: A critical appraisal of the Baltimore renaissance. *Journal of Urban Affairs, 9*(2): 103-123.

Maushard, M. (1993, May 16). Sandtown-Winchester targets crime: Restoring peace, homes are goals. *Baltimore Morning Sun,* p. 5B.

McDougall, H. A. (1993). *Black Baltimore: A new theory of community.* Philadelphia: Temple University Press.

Medoff, P., & Sklar, H. (1994). *Streets of hope: The fall and rise of an urban neighborhood.* Boston: South End Press.

Miripol, A. (1994). Comprehensive neighborhood revitalization: Is resident empowerment taking place? Master's thesis in public policy, John Hopkins University, Baltimore, MD.

Olesker, M. (1993, July 20). The hammering is like music in Sandtown. *Baltimore Morning Sun,* p. 1B.

Pugh, T. (1993, July 22). Renewal project in Baltimore could be beacon for Overtown. *Miami Herald.*

Simmons, M. (1993, September 10). Vacant properties targeted in Sandtown area: Repair or demolition is goal of city grant. *Baltimore Morning Sun,* p. 9B.

Szanton, P. L. (1986). *Baltimore 2000: A choice of futures.* Baltimore, MD: Morris Goldseeker Foundation.

9 The California Mutual Housing Association: Organizational Innovation for Resident-Controlled Affordable Housing

NEAL RICHMAN
ALLAN DAVID HESKIN

Abstract

As providers of affordable housing face dwindling political support and public subsidies, one response may be to explore democratizing housing provision. Three years since its incorporation, the California Mutual Housing Association (CMHA) is carving out a niche in California, providing technical assistance in housing production and operations to municipalities, community development corporations, cooperatives, mutual housing associations, and tenant groups. This case study begins with a review of changing market conditions and development opportunities in Los Angeles. The emphasis then shifts to CMHA's development principles, in part by comparing it with traditional community development corporations. The aims are to elucidate the way CMHA is addressing development needs that previously remained largely unmet and to examine the long-term prospects for a new organization seeking to harness the capacity of housing consumers. CMHA appears to be finding a viable way to serve a growing resident-based constituency for affordable housing. Regardless of the outcome, the experiment provides an opportunity to explore the potential of a form of housing provision, rooted in the practice of democracy within poor communities.

■

The California Mutual Housing Association (CMHA) is an innovation aimed at democratizing the provision of affordable housing. Mutual housing associations are becoming increasingly common, but the statewide, technical assistance character of CMHA makes it unique. In this chapter, we examine the forces that led to the establishment of CMHA and its potential to create new housing organizations and reshape existing ones in ways that are more directly accountable to housing consumers. With centralized, hierarchic forms of nonprofit and public housing provision facing criticism in many countries, the lessons from CMHA may have value in the United States and beyond.[1]

One of the central findings in the 1987 Global Report on Human Settlements was that

> there appears to be no alternative to the one of governmental support for self-determined, self-organized, and self-managed programmes within the framework of settlement-wide action. There can be no guarantees that enabling settlement strategies will work. The problems are too great and the uncertainties too high for such assurances. It does seem certain, however, that they carry the greatest hope and promise. (Habitat, 1987, p. 208)[2]

Although the findings were shaped significantly by the experience of development specialists from the northern hemisphere working in the southern one, the North has been slow to heed its own advice. Even within the Habitat report, chapters are divided, separating development recommendations for the "developed" world from those for the "developing" world. Nearly a decade later, the poorest communities can no longer—if they ever could—be placed neatly along a geographic or temporal development continuum. After decades of intense global capital and residential migration, "third world" enclaves are becoming firmly rooted in "first world" cities, with Los Angeles being an exemplar.

It has been predominantly in the Latin American immigrant communities of Los Angeles, many of which contain the poorest neighborhoods of the city, that resident self-managed programs for development have first taken root with assistance from CMHA. Lessons may perhaps be drawn from CMHA's experience for other world cities seeking to address unmet housing needs through self-managed programs.[3]

After several years of discussion among cooperative housing activists and professionals, tenant organizers, and leaders of community-based nonprofit development corporations, CMHA was incorporated in late 1991 to provide assistance to cooperatives, mutual housing associations, community development corporations, and tenant associations committed to strengthening and expanding not-for-profit, resident-controlled, affordable housing.

In this chapter, we focus on the Southern California portion of the organization where some notable CMHA successes have emerged. CMHA has accelerated the diffusion of housing innovation in a part of the United States that has only recently awakened to the need for affordable housing. CMHA's significance lies in its rapid growth and in the breadth of its technical assistance program. This suggests that the organization is identifying and addressing unmet development needs. The broad scope of activities reflects the diversified CMHA funding base that can help it survive periods of resource drought.[4] On the other hand, the dynamics of intra- and interorganizational relationships are difficult to predict at this early stage.

To examine the potential of CMHA, this analysis begins with a description of recent trends in affordable housing development in Los Angeles against a backdrop of changing market conditions. CMHA's development approach is then compared to that of CDCs in Southern California to explain the niche that this resident-controlled technical assistance organization is seeking to carve out for itself. The conclusion evaluates CMHA's long-term prospects.

■ Nonprofit Housing Provision in Southern California

Late in coming to Southern California, the nonprofit housing movement has grown over the past seven years to include more than 30 community-based housing organizations citywide (Squier & Associates, 1990). They have rapidly responded to a shortage of affordable housing. Yet, new challenges are arising that will test the ability of these CDCs to adapt to new neighborhood and fiscal conditions.

With only a few notable exceptions, CDCs in Los Angeles did not grow out of a long tradition of grassroots neighborhood advocacy. Rather, CDC expansion has been based in recruiting and training leadership, drawn from local social service organizations, ethnic and cultural associations, and churches. This leadership had been identified and groomed by the Local Initiatives Support Corporation (LISC), a national nonprofit intermediary. LISC brought the tax credit partnership program to Southern California in the 1980s, promulgating it through sponsorship of CDC training programs and funding new ventures through its offshoot, the California Equity Fund (Heskin, 1992). Although LISC introduced the techniques used in public-private housing partnerships to Los Angeles, the intermediary could not transfer the years of neighborhood advocacy experience embedded within community organizations in the older cities in the East.

The structure of the tax credit program led to a focus on the construction of new rental housing as the primary component of a neighborhood-based development strategy. CDCs in Southern California, like their counterparts around the country, match a combination of federal, state, and local funds with private investment from financial institutions and investors seeking tax credits to produce this housing (Hecht, 1994). Organizational funding is secured by development fees that are incorporated into project budgets to cover the cost of assembling these complex public-private financial partnerships. The complexity of assembling these public-private partnerships can overwhelm community development corporations and may result in displacing other important organizational objectives (Richman, 1996).

The political legitimacy of these nonprofit housing providers tends to be based in neighborhood leaders. The provision of on-site services, such as child care and job training, sometimes called "Housing Plus," serves as the primary "community development" component of CDC activities.

■ Los Angeles Undergoing Transformation

Los Angeles has become an international symbol of the urban crisis in advanced capitalist societies. The wall dividing housing opportunities by class is increasingly insurmountable.[5] Only five years ago, the affordable housing crisis was widely perceived as a problem of housing shortage. Although the deficit of these dwelling units has grown in absolute numbers, overlapping this trend have been wide-scale disinvestment, declining property values, residential deterioration, and rising vacancy rates. "Rust-belt" housing markets have arrived in the very capitol of the "sun-belt." The City Housing Department's shift to funding purchase and rehabilitation rather than new construction grows out of this change. Yet, it is taking some time for the institutions of housing provision—public, private, and nonprofit—to "retool their machinery."

Housing studies based on the 1990 census show that Los Angeles, compared to other U.S. cities, has an extremely high percentage of rent-burdened tenants and an extremely low proportion of renters who are economically capable of homeownership (Myers, 1993).[6] Especially in low-income neighborhoods, ownership rates are very low, with as much as 75% of the single family houses in some neighborhoods in the hands of absentee landlords. Overcrowding doubled since 1980. In some instances, more than ten people crowded together in substandard two-room flats, lacking plumbing and cooking facilities. Rodent infestation, lead poisoning, and death by fire—all became more prevalent as threats to health and safety.

With the enormous need for decent, affordable housing in the region, great pressure is placed on the limited subsidized stock, much of which is in private hands. Approximately 15,000 dwelling units, the largest concentration in the United States, are located in "expiring use" or "at risk" projects. These are developments in which federal rent regulatory restrictions are about to expire, permitting owners the choice to opt out of subsidy programs and displace low-income households in effect by raising rents to market levels (Clay, 1987).

Struggling through a period of economic decline in the wake of the 1992 riots, less than two years later, Los Angeles was rocked by another upheaval—this time seismic. More than 65,000 residential structures, containing more than 300,000 dwelling units, were damaged, resulting in a total loss of approximately $1.5 billion in the city's housing stock. Nearly 20,000 dwelling units had to be vacated, 90% of these multifamily units.[7] In some instances, these neighborhood concentrations of unoccupied housing became "ghost towns," quickly invaded by the homeless, drug dealers, and gang members.

Yet surprisingly, this catastrophe-induced housing shortage did little to bolster property values in other parts of the city or to increase overall residential occupancy. Rather, the housing situation has only worsened to a depression-like combination of overcrowding and growing homelessness in the 1980s with high vacancy rates.[8] The income of renters in the city's oldest housing (stabilized rentals) has dropped 35% since 1990 (Gordon, 1994) as the housing market in Los Angeles is increasingly characterized by disinvestment, residential foreclosures, and housing abandonment. According to the city building department, the number of abandoned properties increased by 28% over the past two years alone, now totaling 1,381 (Gordon, 1994).

Shifting the Direction of City Housing Policy

The changing conditions in Los Angeles has led the city's Housing Department to change its policy away from an emphasis on new construction to meet affordable housing shortages to an emphasis on purchase and rehabilitation of existing underutilized and deteriorating housing. In part, this change has been dictated by simple mathematics. Development costs of new housing have been high. In Los Angeles, the cost of building one new multifamily unit on the lowest-cost land equals approximately $150,000, making very high subsidies essential. In contrast, CMHA is showing that rehabilitated housing in downtown can be produced at approximately half the cost of new construction.

The shift in city policy has disconcerted many CDCs that previously had relied heavily on new construction. Rehabilitation was typically only undertaken in large empty buildings. As a result, very few of the CDCs in Los Angeles have had experience in dealing with the existing occupants in residential structures. Abandonment of smaller structures predominates this and increasing problems in partially occupied buildings, reshaping the nature of CDC development opportunities. For example, a recent innovation, attracting interest from CDCs and funding from the city, is a development plan involving the acquisition of low-density scattered site parcels—primarily "nuisance" properties where drug sales, prostitution, extreme physical deterioration, or other signs of abandonment are apparent. These parcels—containing two to four dwellings per site—will be developed as one project by the CDC and leased to the resident-based cooperative for the 15-year period of the tax credit partnership.

The California Mutual Housing Association

In contrast to CDCs, CMHA relies on a political base of lower-income resident groups, organized around housing issues, rather than a base that is defined geographically, typically at the neighborhood level. Being statewide, the organization can respond to need and opportunity wherever they occur. Building from active tenants in place, CMHA can respond to residents in HUD, public, and distressed private housing. It can also help provide forms of resident control that are more consistent with community development objectives than a continuing CDC reliance on rental housing. When CDCs are interested in pursuing forms of resident management, CMHA can help the groups harness grassroots energy and commitment.

Important struggles around resident-controlled housing have emerged in Los Angeles over many years (Haas & Heskin, 1980; Heskin, 1991). Nevertheless, CMHA is the first technical assistance center established to support these grassroots efforts, directly accountable to those it serves.

Since hiring the first staff member 18 months ago, the Southern California office has grown rapidly and operates on a budget of approximately $300,000 although the potential income of the organization is much higher. The income sources of CMHA in Southern California are quite diverse, but about half of the budget funds will be available through federal technical assistance contracts, one to support CDCs embarking on resident-controlled housing and others to support the development efforts of resident associations in HUD subsidized at-risk buildings. Given the Republican sweep of Congress and new proposals from the Clinton administration, these federal funds are likely to dwindle within two to three years. The long-term impact of CMHA will depend on how it manages this critical period.

Currently in Southern California, more than 20 developments—representing more than 4,000 units—have received services from CMHA. These buildings have a combined gross annual income of more than $24 million; hence, the development efforts of CMHA could make an important contribution to local economies and create a stable client base for ongoing operations. Moreover, in its short history, CMHA has helped to develop

- the first HUD "expiring use" rental conversion to resident-controlled nonprofit housing in Southern California;
- the first acquisition and rehabilitation project by renters of privately owned slum housing in Los Angeles; and
- the first limited equity cooperative conversion projects in the City and County of San Diego.

It is important to note that CMHA is an association of autonomous groups rather than a large holder of property. This structure prevents the rigidity of such large organizations, although it may result in CMHA attaining less economic and political power than its large predecessors in Northern and Western Europe. To compensate for this, CMHA has taken a broad definition of resident control, including limited equity cooperatives, mutual housing associations, and tenant associations seeking a strong role in the operation of the rental housing they occupy. This broad definition of resident control has expanded the potential orbit of the organization. Most notably, this structure can link tenants in both public and private housing with CDCs and cooperatives to broaden their collective power.

CMHA's Development Principles

CMHA activity is governed by core principles that have evolved from the experience of the organization's leadership and from its pragmatic experience of development practice. These principles differentiate CMHA from its more mainstream nonprofit development counterparts.

Principle 1: "Don't Own, Organize"

Increasingly, CDCs have accepted a development model aimed at establishing an asset base of rental housing. Accordingly, control of real estate ultimately contributes to the parent organization, as properties are used for collateral, for refinancing, and ultimately to generate long-term income. The unfortunate reality for CDCs is that real estate holdings often cost more to responsibly manage and operate than they generate for the parent organization, especially during the project start-up years. In the United States, property management in low-income communities has historically been underfunded. Moreover, when the nonprofit is acting as a general partner in a tax credit syndication, monitoring and reporting responsibilities are burdensome with costs often not fully compensated by fees. Finally, any remaining cash flow tends to be gobbled up by government funders seeking cost recovery of their no-interest or low-interest loans.

Other problems are also inherent in this model. Property management companies tend not to receive adequate oversight by the CDC staff, and resident complaints are increasingly finding their way to elected officials. Rather than rely on their tenants to oversee housing services, CDCs in Los Angeles have begun to hire asset managers. Furthermore, the attempts by CDCs to increase the size of their holdings are often attacked by opponents as self-serving. Sometimes, these opponents are renting from the CDC, concerned that there is already insufficient attention from the parent corporation.

CMHA does not approach its development work with the goal of establishing an asset base. Rather, the objective is to work with and create local housing organizations that provide a support network for resident-controlled housing. This network is seen as the long-term base that will sustain CMHA and its services. CMHA is almost entirely supported by billing hours to particular projects, for either development activities or assisting existing cooperatives. Unlike for CDCs, development fees have been a relatively minor part of CMHA's budget.

Although the CDC movement is new in Southern California, groups are considerably turf-conscious. As a statewide resident-controlled development organization, CMHA can work with local CDCs that want to expand skills in resident ownership forms, or CMHA can go into another's turf to work with residents of a building, if the CDC is not interested in the project. If there is no CDC or no CDC interest in resident-controlled housing, then CMHA can help community leadership create a local MHA to further the movement. In this case, CMHA acts as staff to the new local MHA, until it is able to sustain its own local staff. The real estate activity development is a vehicle for organizing residents and community leadership, not an end in itself.

Principle 2: "Begin With the People, Not With the Building"

Typically, when Los Angeles-based CDCs seek residential properties for acquisition and rehabilitation, they work with local real estate brokers to identify a "buy" that can be packaged using tax credit funding. Ideally, properties are vacant to avoid the difficulties and costs associated with relocation. CMHA, in contrast, seeks first to develop where there are active resident associations, and second, where a development opportunity lends itself to producing resident-controlled housing.

One byproduct of this approach is a change in the local politics of development. Many CDCs have had to face the "Not In My Back Yard" (NIMBY) phenomenon, even in lower-income neighborhoods. This is much less likely with resident-controlled housing. Working with tenants who have already battled for the right to buy means that a development program often follows political conflicts. The CMHA approach changes the debate around affordable housing projects. It is one thing for affordable housing opponents to battle against a project for an unidentified population; it is quite another for them to oppose the interests of existing residents. CMHA has also found that resident-controlled housing, also known as "ownership," is more acceptable to neighbors and more politically conservative representatives.

Starting with the residents has linked the tenant movement more directly into project-specific development activity. Some tenant organizers believe that only adversarial organizing has political potential, but CMHA's experience is that most tenants would rather assume authority than expend energies forcing a landlord to make needed repairs. Tenant organizing can focus productively on gaining control of the real estate asset and establishing and overseeing a reasonable operating budget.

Bringing housing residents directly into development decision making can yield powerful practical results. For instance, when the City of Los Angeles Housing Department began to require the addition of certain expensive and time-delaying design features to enhance the marketability of rehabilitated slum housing, the developer stepped aside and let the residents take over. The residents themselves made it clear that they did not share the city's architectural vision. The direct dialogue provided opportunities for cultural differences and real-life experience to enter into the architectural process. The city's architect was willing and able to fight with the professional developer. He was not prepared to have the same fight with the residents.

A resident-centered development program means that CMHA staff has not had to market its services. The growth of the organization, especially in Los Angeles, has been through word of mouth. Tenant groups contact other tenant groups and cooperatives contact cooperatives. Each success builds this network of references and enlarges the potential market for CMHA services.

Principle 3: "Let the People Decide"

CDCs constellate around development deals, typically those involving property acquisition, financing, and construction activities. For CMHA, the emphasis is on social development, especially training resident associations in property operations and management. This is one of the reasons CMHA takes such a broad definition of the term *resident-controlled housing*.

To approach development from this resident-based perspective requires that CMHA have staff skilled on multiple levels, as organizers as well as real estate developers. Rather than simply getting through development checklists, CMHA staff must frame development problems as opportunities for organizational capacity building. They must repeatedly be asking the question, "How can decisions be made in ways that enhance a tenant organization's long-term strength?"

A major challenge is that residents may seek to draw CMHA staff into internal conflicts as adjudicator or ally. CMHA must be careful in supporting democratic processes rather than stepping in with "the answer" or becoming aligned with one faction or another. Awareness of this problem has thus far meant that CMHA has survived the bitter fights that can occur in resident-controlled housing organizations. CMHA staff has learned that a reasonable resolution will ultimately be reached by the people.

Through a process of structuring and resolving real estate decisions, resident groups learn how to increase control over their living environment. In CMHA's work with a resident group in slum housing, one major task was redesigning overcrowded single- and double-room flats to accommodate the existing families. The residents designed an internal relocation plan and divided up vacated space to accommodate those with the greatest need. The work required collectively resolving issues of equity—who has priority for living space, on which floor, near which household. Perhaps as important, the residents began to see the deteriorated building that both sheltered and victimized them as theirs, as malleable, as reparable, and as something that they could collectively shape and improve. Under federal law, each tenant household was entitled to a payment of $300 to $400 to cover the moving costs in relocating temporarily during construction. All 22 households waived their rights, handling the move collectively and contributing the unused funds to supplement the rehabilitation budget.

Principle 4: "The Marriage Is More Important Than the Wedding"

Too often in CDC-sponsored residential development, property management is an afterthought, what one does after the real development—financial packaging and construction—is completed. When this occurs, CDCs mirror the activities of market rate developers who build properties with a short-term profit margin.

The CMHA approach turns this view on its head; the point of property development is not development and the developer fee but rather the opportunity to control an operating budget. These funds are viewed like an endowment for democracy. Project operating budgets can generate a steady stream of resources each month for further investment in board training and other services.

Development consultants typically swarm around a project when it is in the planning and construction phases then abandon the residents when the resources have been raised and expended. In co-op training, residents are reminded that all these development activities are just like preparing for a wedding ceremony. The residents must direct the florist and the caterer, not vice versa. But most important, they must remember not to mix up the wedding event with the importance of building a strong relationship for the many years of marriage. Just as couples can best live together with each other over the years, if they have open and respectful ways of communicating, so too can cooperatives. Marriages are also better able to survive, if income is sufficient to cover bills, hence the importance of proper budgeting and fiscal management to a co-op.

Over time, CMHA plans to charge established resident-controlled housing development fees for providing basic ongoing services. CMHA can then use that contract-based income stream to cover a portion of its operating budget. For example, resident-controlled housing projects in Southern California can contract for a "corporate secretary" based on a fee of 1% of gross receipts. The services covered will include maintaining a meeting minute book, overseeing elections, making sure that procedures follow the bylaws, and other organizational development functions not provided by private property management companies.

One reason that many CDCs claim that they do not want resident control is that over time co-ops may lose their democratic orientation and, in some instances, control of the affordable housing. Co-ops being left on their own without support can be taken over by internal cliques or hand over control to their management company. One disreputable management company in Los Angeles did not even pay property taxes, resulting in titles held by two cooperatives placed in jeopardy. The corporate secretary role is one mechanism to ensure ongoing technical assistance.

To help prevent mismanagement and abuse of rules, CMHA has recommended that the resident-ownership structure include two other components: first, the inclusion of at least one community representative on the project board of directors to provide access to resources in the surrounding community and to guard against isolation, and, second, an organizational audit performed each year in addition to the normal financial audit, with the purpose of identifying organizational problems and designing an intervention strategy before they get out of hand. The corporate secretary, outside member, and audit have overcome much of the resistance of lenders to financing resident-controlled housing in California. Lender acceptance of the model and CMHA is indicated by the Savings Association Mortgage Corporation (SAMCO), a coalition of savings and loans organized for socially responsible investment. SAMCO has made the involvement of CMHA a requirement in its financing of resident-controlled housing.

Resident oversight of housing activities provides long-term framework for community development. Typically, community organizations know what needs to be accomplished in their neighborhood but lack the necessary resources and administrative support. These groups must try to pressure government, banks, and others to do what is needed. But, if residents manage the property, they can apply their own experience-based understanding of the property and the neighborhood toward solving existing problems. For example, instead of paying (again) for graffiti removal, a resident group can use operating funds to start a small-scale after-school program for youth and evaluate the results.

Both CMHA and local CDCs seek community development. But their approach is different. The CDC seeks to achieve its mission through projects and programs that it brings to the community. CMHA seeks the same result by nurturing a resident-controlled institution whose members have the confidence to design their own programs for the area.

Principle 5: "Rely on People-to-People 'Peer-Based' Technical Assistance"

With the growing complexity of CDC residential development projects has come increased reliance and focus on professional staff and consultants. This form of organizing is far from the roots of CDCs that emerged from block clubs and neighborhood associations. Moreover, with the increasing technical complexity of development, there is growing difficulty in hiring local residents to perform CDC functions. By the old route, one moved up from community organizer to housing development staff. Today, the minimum qualifications for a housing packager typically require a graduate degree plus experience.

CMHA seeks to break down the division between technical staff and resident leadership. The greatest knowledge of running resident-controlled housing comes from having done it. This is neither textbook knowledge, nor computer knowledge, nor something one learns in graduate school. Rather, operating resident-controlled housing requires very specific and context-based understanding, drawing upon knowledge of a particular community and its buildings.

Beyond the ability to read management financial reports or interpret bylaw provisions comes the requirement that people have the democratic process skills to resolve thorny value-based problems in ways that build group cohesion and fairness. Many decision-making issues have important moral components that cannot be simplified into formulaic technical problem-solving. These decisions involve the creation and interpretation of policies and procedures for admitting new members, revoking membership rights, establishing priorities for repairs, and so forth. Often these decisions are not clear cut but instead require reliance on the judgment of those most involved and most knowledgeable of the particulars.

Sometimes, those with the most knowledge can be found within the resident association, but often much can be learned from those who have struggled through similar issues in other settings. In cooperative development training, it is very important to teach the residents how to identify resource people within and outside their group.

CMHA staff and board representatives help in this networking function. In addition, in the training process, residents meet CMHA co-op leaders who can be contacted for follow-up. For example, under CMHA's contract with HUD for technical services, Spanish-speaking resident leaders from Los Angeles were brought down to San Diego county to provide training. CMHA's strength lies in these direct people-to-people exchanges. CMHA will not work with a CDC or a tenants association unless they agree to make their acquired skills available to others.

The central mechanism for strengthening the interorganizational CMHA community is through the week-long Twin Pines Summer Housing Institute at Co-op Camp Sierra. CMHA is the only development corporation in the state with its own folk school for continuing education. With informal training sessions and informal interactions, Co-op Camp Sierra helps break down divisions between technical staff and residents in a retreat setting. Participants bringing their entire families begin to establish long-term friendships as well as broad co-op networks.

■ What Can Go Wrong?

Cooperatives fail when internal conflicts or funding difficulties undercut the initiatives; conversely, they also fail when, by institutionalizing their accomplishments, they do not maintain an energetic commitment to democratic process and grassroots innovation. In this connection, five questions are critical to CMHA's long-range potential.

1. Can CMHA use short-lived federal funds for providing technical assistance to CDCs and to residents in HUD-subsidized housing in a manner that strengthens relations with these groups, ideally resulting in new markets for CMHA services? In addition to being driven toward purchase and rehabilitation of occupied buildings by city housing policy, other forces may be bringing the CDCs closer to CMHA. Requests from CDCs for technical assistance from CMHA suggest a growing market for services; however, as subsidies dry up, these may become a lower priority. CMHA should seek to demonstrate the cost effectiveness of replacing paid "asset managers" with the volunteer efforts of those who know the properties best—the residents. No one is better suited to be the eyes and the ears of the nonprofit owner than those located *in situ.*

With regard to residents in HUD expiring-use buildings, federal funding may be canceled and the political problem of displacement remains. In Los Angeles, active tenant associations will likely turn to local government for support, including funding for CMHA technical assistance. During fiscal austerity, CDCs proposing to build new housing may have difficulty competing for resources with the united voices of subsidized renters facing the loss of their homes.

2. Once established, will the independently owned, resident-based housing associations support and contribute to CMHA and its network of resident associations? Or conversely, engage in unnecessary conflict and competition? This is hard to predict. Co-op Camp and the Twin Pines Housing Institute hold the promise of providing the social glue that binds these disparate groups together with the cooperatively controlled technical assistance providers. On the other hand, there are numerous cases where the parent organization was rejected by resident-controlled offsprings.

3. Will CMHA be able to market and provide a level of support services that established cooperatives require but often do not provide by themselves? CMHA has had some success at negotiating agreements for "corporate secretary" services with existing cooperatives. A fee of 1% of gross effective receipts has been tied to the provision of a specified number of CMHA staffing hours. In a number of cases, property management companies have agreed to reduce their fee by .05%, because CMHA services can smooth overall operations. As a result, for a minimal fee cooperatives get access to top rate housing experts. These standard agreements establish a base rate and may lead to additional billing for services as required by the cooperatives.

The significance of this revenue will depend on the number of new resident-controlled developments, CMHA can spawn. As noted, the Savings Association and Mortgage Corporation—the primary community reinvestment vehicle of the savings and loan industry in the western region of the United States—requires that resident-controlled projects in California establish training and service relationships with CMHA in its loan document regulatory agreements.

4. Can CMHA avoid a centralization of power in the paid staff where internal organizational needs displace more forward-looking goals? One can find instances where grassroots development initiatives were coopted by more professional managerial organization to meet the demands of staff or funding agencies (Richman, 1995). The evolution of the Amalgamated Clothing Workers Cooperatives in the Bronx into Abraham Kazan's regional United Housing Foundation followed such a path, ending with the Co-op city debacle (Leavitt, 1995).

Centralization of power is difficult to predict. The support for CMHA staff from residents can buoy their energy, but requirements for night meetings and community process can take their toll. Perhaps, the board structure with 70% control held by existing and emerging resident-based housing organizations provides the best assurance for ongoing accountability.

5. Will CMHA and its network of resident-based housing groups allocate surplus resources to expand "ownership" opportunities for low-income renters or serve primarily those in resident-controlled housing? How will the right balance be reached? The CMHA Southern California office has emphasized serving new groups over existing ones, in part because the existing market for CMHA services is not large enough to ensure its survival. Again, the board of directors has been structured to have representation from organized tenant groups seeking resident-controlled housing, so as to reach out and serve emerging development needs.

CMHA combines local legitimacy with centralized services. CMHA provides direct accountability to local constituents. Like CDCs, it can also harness grassroots energy and commitment. Yet, as a secondary cooperative—a co-op of co-ops—CMHA functions on a statewide and regional basis with potential economies of operation. Each local housing organization need not carry its own development staff during times of funding drought, if development support is available via an umbrella group.

CMHA will be tested in the years ahead. Regardless of the outcome, the experiment represents an important opportunity to explore the potential of a form of housing provision, rooted in the practice of democracy within poor and working-class communities.

NOTES

1. The spread of community development corporations has been heralded as a democratizing force in affordable housing provision in the United States (Harloe & Martens, 1990; Mayer, 1988; Peirce & Steinbach, 1987; Pickman, Benson, Centerman, & Mittle, 1986; Vidal, 1992) Yet, despite many important contributions by these organizations to their neighborhoods, CDCs have not escaped criticism that they lack grassroots political support. Typical of the critics is Richard Taub (1990), based in Chicago, who evaluated the community development work of CDCs:

> People in communities with CDCs seldom perceive the output of the corporations as the product of their own companies working for them. . . . In fact, community residents often have no idea of the Herculean efforts underway on their behalf. The CDC is a "they," not terribly different from other things that impinge on their lives. (pp. 7-8)

A similar conclusion is reached by Dennis Keating (1989) about Cleveland and by Harold DeRienzo (1994), writing about his experience as director of the Banana Kelly Community Improvement Association in New York. Also, based in the New York region, Susan Fainstein (1990) described community planning in New York as increasingly a matter of investment targeting aimed at pleasing powerful public and private "partners."

2. The 1987 Habitat document argues strongly for optimum participation by residents in self-managed programs. The description of the areas of participation reads much like a CMHA work plan:

- *Participation in planning*—in the definition of objectives, strategies, and priorities.
- *Participation in programming and budgeting*—guaranteeing the effective employment of resources to fulfil objectives.
- *Participation in implementation*—creating responsibility for maintenance and management.
- *Participation in operational activities*—securing more cost-effective and efficient maintenance and management. (Habitat, 1987, pp. 104-105)

3. The Megacities project has produced research about the international transfer of community-based innovations across North-South boundaries. See, for example, Perlman, 1990; Badshah and Lazar, 1995.

4. Examples include development services for tenant associations purchasing four HUD-subsidized prepayment developments; development services for acquisition and rehabilitation of abandoned slum buildings by its residents in downtown Los Angeles; technical assistance to seven CDCs throughout Southern California; creation of four local mutual housing associations—organized to operate resident-controlled housing; acting as corporate secretary—ensuring corporate compliance and providing staff support—for three existing limited equity cooperatives; leadership training for residents in three public housing developments; training for municipal staff members in two cities; and advocacy for a number of resident organizations fighting their landlords and seeking opportunities to purchase.

5. The roots of this polarization lie in part with the massive plant closings and other economic dislocations that occurred over the past two decades, decimating the number of blue-collar and middle-class jobs. With more recent cutbacks in defense expenditures, another source of higher wage employment fell in a region very dependent on military-related industries.

6. The executive director of the Neighborhood Housing Service of Los Angeles reports that even with the use of all forms of creative financial leveraging, only one of 30 applicants for homeownership have the incomes to qualify for purchase in low-income neighborhoods.

7. Another 28,000 dwelling units face a similar fate. They are at risk of being lost from the city's housing stock because of the expected owner's difficulty in financing the high cost of repair work. The boom housing decade of the 1980s resulted in many apartment buildings being overfinanced. Collusion between bankers, appraisers, and owners—whether illegal or unethical—provided mechanisms to churn properties and use inflated values to build mini-real estate empires. The earthquake brought down more than just buildings, it punctured a distorted speculative bubble.

8. The official vacancy rate is 8%.

REFERENCES

Badshah, A., & Lazar, R. (1995). Sharing approaches that work: Transfer and adaptation of urban innovations. *Cooperation South.* New York: United Nations Development Program.

Clay, P. L. (1987). *At risk of loss: The endangered future of low-income rental housing resources.* Washington, DC: Neighborhood Reinvestment Corporation.

DeRienzo, H. (1994, December). Managing the crisis. *City Limits, 19:* 10-25.

Fainstein, S. S. (1990). Neighborhood planning: Limits and potentials. In N. Carmon (Ed.), *Neighborhood policy and programs: Past and present.* New York: St. Martin's.

Gordon, L. (1994, December 10). Fighting to reclaim blighted buildings. *Los Angeles Times,* pp. A1, A26-A27.

Haas, G., & Heskin, A. D. (1980). Community struggles in Los Angeles. *International Journal of Urban and Regional Research,* p. 5.

Habitat—United Nations Center for Human Settlements. (1987). *Global report on human settlements.* Oxford, UK: Oxford University Press.

Harloe, M., & Martens, M. (1990). *New ideas for housing: The experience of three countries.* London: Shelter.

Hecht, B. L. (1994). *Developing affordable housing: A practical guide for nonprofit organizations.* New York: John Wiley.

Heskin, A. D. (1991). *The struggle for community.* Boulder, CO: Westview.

Heskin, A. D. (1992, July). *Housing reform in Los Angeles.* Annual Conference of the Collegiate Schools of Planning and the Association of European Schools of Planning: Planning transatlantic global and local problems, Oxford, UK.

Keating, W. D. (1989, February, March, April). The emergence of community development corporations: Their impact on housing and neighborhoods. *Shelterforce.*

Leavitt, J. (1995). The interrelated history of cooperatives and public housing from the thirties to the fifties. In A. Heskin & J. Leavitt (Eds.), *The hidden history of cooperatives.* Davis, CA: Center for Cooperatives, University of California.

Mayer, N. S. (1988). *The role of nonprofits in renewed federal housing efforts.* Boston: MIT Center for Real Estate Development, Housing Policy Project.

Myers, D. (1993). *How are we housed? Affordability, overcrowding, and achievement of homeownership in Los Angeles from 1980 to 1990.* City of Los Angeles, Housing Department.

Peirce, N. R., & Steinbach, C. F. (1987). *Corrective capitalism: The rise of America's community development corporations.* New York: Ford Foundation.

Perlman, J. E. (1990, February). A dual strategy for deliberate social change in cities. *Cities, 7*(1).

Pickman, J., Benson, R. F., Centerman, M., & Mittle, R. N. (1986). *Producing lower-income housing: Local initiatives.* Washington, DC: Bureau of National Affairs.

Richman, N. (1995). From worker cooperatives to social housing: The transformation of the Third Sector in Denmark. In A. Heskin & J. Leavitt (Eds.), *The hidden history of housing cooperatives.* Davis, CA: Center for Cooperatives, University of California.

Richman, N. (1996). Beyond public/private partnerships: CDCs and affordable housing provision in the USA. In A. Twelvetrees (Ed.), *Rhetoric or reality? Community economic development in the U.S. and U.K.* Wales, England: Pluto Press and the Community Development Foundation.

Squier, G., & Associates. (1990). *Nurturing community development.* Los Angeles: California Community Foundation.

Taub, R. P. (1990). *Nuance and meaning in community development.* New York: Community Development Research Center, New School for Social Research.

Vidal, A. (1992). *Rebuilding communities: A national study of urban community development corporations.* New York: Community Development Research Center, New School for Social Research.

10 Housing Trust Funds: A New Approach to Funding Affordable Housing

MARY E. BROOKS

Abstract

This chapter provides an overview of housing trust funds as they have developed in this country. Housing trust funds provide a dedicated ongoing source of local revenue to support projects that address the housing needs of low- and very low-income residents. There are now more than 100 such funds being implemented in cities, counties, and states. Broad parameters for what characterizes these funds are given, including what is meant by a dedicated source of revenue, how the funds are administered, what kind of program requirements are developed, and the kind of partnerships that are created as projects are funded.

Descriptions are provided of how three local housing trust funds were developed. These include the City of Chicago's Low-Income Housing Trust Fund, Boulder's Community Housing Assistance Program, and San Diego's Housing Trust Fund. Each example provides information about how the fund was created under circumstances unique to that city, how the fund presently operates, and what it has been able to accomplish.

■

Although the concept of dedicating a specific revenue source to a specific purpose is not new, its application to housing only started to catch the attention of housing advocates, elected officials, and housing planners in the 1980s. Since then, the number of housing trust funds has grown rapidly. There are now more than 100 housing trust funds in cities, counties, and states throughout the country. As few as seven existed only 10 years ago (Brooks, 1994; see Figure 10.1 and Table 10.1).

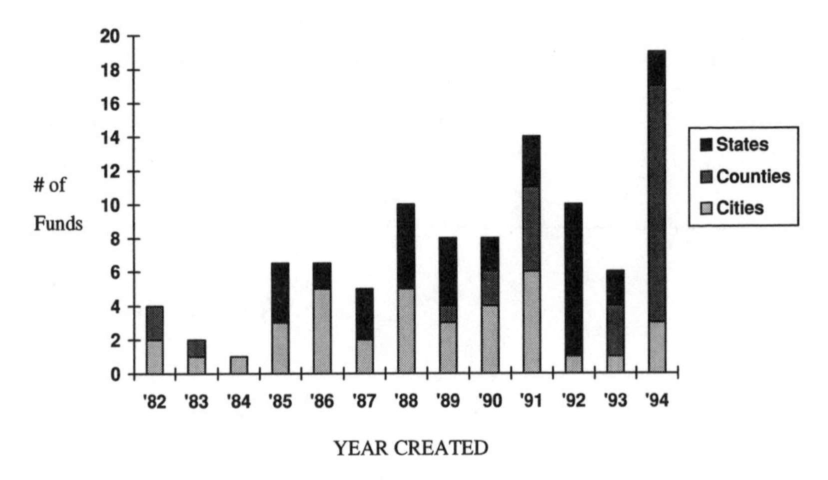

Figure 10.1. Housing Trust Funds Created by Year

Precedent for the concept of dedicating public revenues to a specific purpose exists in several forms. Interest on lawyer trust accounts (IOLTA) is collected in most states throughout the country to support the legal services program. IOLTA programs collect the revenue from deposits made in advance of legal services by those who can afford legal representation. When they are not placed into individual interest-bearing accounts, the interest is aggregated on a statewide basis. The revenue is then committed to provide legal services to those who cannot afford their own legal representation.

Moreover, earmarking taxes is not uncommon. The National Conference of State Legislatures reports that in fiscal year 1988, almost $50 billion or nearly 23% of state tax collections were earmarked for specific state programs. The most common kind of earmarking is dedicating motor fuel taxes to highway and other transportation programs (Fabricius & Snell, 1990, p. 7).

Property tax revenues at the local level have a long-standing affiliation with the support of school systems. Identifying similar revenues that could be dedicated to support housing for those who cannot afford it is thus an extension of existing fiscal practices. Two of the earliest sources for housing trust funds were the real estate transfer tax (a tax paid at the time real estate is transferred) and linkage fees (fees paid by commercial and industrial developments to offset the impact of additional employees on the local housing supply).

TABLE 10.1 Housing Trust Funds (1995)

City	*County*	*State*
Alexandria, VA	Allegheny County, PA	Arizona
Ann Arbor, MI	Berks County, PA	California
Bellevue, WA	Bucks County, PA	Connecticut
Berkeley, CA	Centre County, PA	Delaware
Bloomington, IN	Chester County, PA	
Boston, MA	Crawford County, PA	Florida
Boston, Local 26, MA	Dade County, FL	Georgia
Boulder, CO	Dade County, FL	Hawaii
Burlington, VT	Dauphin County, PA	Idaho
Cambridge, MA	Dayton/Montg. County, OH	Illinois
Chicago, IL	Fairfax County, VA	Indiana
Cleveland, OH	Howard County, MD	Iowa
Cupertino, CA	Indiana County, PA	
Denver, CO	Jackson County, MO	Kansas
Duluth, MN	King County, WA	Kentucky
Hartford, CT	Lackawanna County, PA	Kentucky
Knoxville, TN	Lancaster County, PA	Maine
Los Angeles, CCW, CA	Lebanon County, PA	Maryland
Manassas, VA	Montgomery County, MD	
Memphis, TN	Montgomery County, PA	Minnesota
Menlo Park, CA	Napa County, CA	Missouri
Miami, FL	Sacramento County, CA	Missouri
Morgan Hill, CA	St. Charles County, MO	Nebraska
Modesto, CA	St. Louis County, MO	Nebraska
New York City, BPC, NY	Santa Clara County, CA	Nevada
Palo Alto, CA	Summit County, OH	Nevada
Phoenix, AZ	Toledo/Lucas County, OH	New Hampshire
Pittsburgh, PA	Westmoreland County, PA	New Jersey
Sacramento, CA	Warren County, PA	North Carolina
Saint Paul, MN	York County, PA	Ohio
Salt Lake City, UT		Oregon
San Antonio, TX		Oregon
San Diego, CA		Rhode Island
San Francisco, CA		South Carolina
San Francisco, CA		Tennessee
Santa Monica, CA		Texas
Seattle, WA		Utah
Seattle, WA		Vermont
Washington, DC		Washington
West Hollywood, CA		Wisconsin

NOTE: Double listing implies more than one housing trust fund.

The notion of dedicating revenue to provide an ongoing source of support for a specific purpose is suitable particularly to the provision of housing for the lowest-income households. First, it demonstrates a long-term commitment on the part of elected officials to a housing supply that is affordable for all residents of the community. Second, in today's housing market, developing housing for low-income households typically requires packaging seven or eight different sources of financing (Hebert, Heintz, Baron, Kay, & Wallace, 1993). Projects can take years from conception to construction. Having a reliable source of funding to contribute can make an otherwise impossible project doable. Third, a relatively reliable source of funding enables units of government to plan how best to address the most immediate housing problems and to be flexible, as priorities may need to shift from year to year. Fourth, knowing that funding is forthcoming each year to support housing programs allows housing trust funds to utilize a portion of its available funds for projects that may be more risky than normally considered desirable; to work with organizations and developers that may not have a large portfolio of successful projects but are willing to undertake difficult projects serving special populations or neighborhoods; and to engage in long-range planning, such as providing capacity building funds to nonprofit development organizations.

Housing trust funds began to appear at a time when the housing crisis was escalating. Housing needs of the poor were growing, and federal assistance was all but disappearing. According to the Center on Budget and Policy Priorities and the Low-Income Housing Information Service, between 1978 and 1985, the number of poor households rose 25%, from 10.5 million households in 1978 to 13.3 million in 1985 (Leonard, Dolbeare, & Lazere, 1989, p. xiv). In addition, those who were poor grew poorer. In 1978, the typical poor family had an income that fell $3,362 below the poverty line. By 1985, its income fell $3,999 below the poverty line—further below than in any year since 1959, when such data were first collected. Although the number of poor households increased, the number of units renting for $250 or less declined. There were 9.7 million such units in 1970, but only 7.9 million in 1985. This represented a loss of 1.8 million low-rent units from the housing stock, a 19% decline (Leonard et al., 1989, p. xiv).

The rise in housing needs of the poor during this period was exacerbated by the absence of adequate government assistance. Fewer than one in three poor renter households received any kind of federal, state, or local rent subsidy or lived in public housing in 1987. Unlike other safety net programs, the poor are not legally entitled to housing assistance, even if they meet all of the eligibility criteria. Rather, the number of households served each year is determined by the level of funding appropriated by the Congress. From fiscal year 1977 through fiscal year 1980, HUD made commitments to provide federal rental assistance to an average of 316,000 additional households per year. From fiscal year 1981 through fiscal year 1988, however, the number of housing commitments to serve additional households dropped sharply to an average of 82,000 per year. The number of additional low-income renters being provided housing assistance each year fell by nearly three fourths (Leonard et al., 1989, p. xiv). The idea of finding some way to provide funding for low-income housing that was relatively permanent, using local sources of revenue, was a natural response to a crisis made worse by an unresponsive federal government. Once a few models became available, between 1980 and 1990, the number of housing trust funds in cities, counties, and states increased 25 times (Brooks, 1989).

How Does a Typical Housing Trust Fund Work?

Housing trust funds now operating around the country actually do not look very much alike in many ways. But most of them do have some common characteristics that not only enable them to be classified as housing trust funds but also give help to define how these funds work.

1. A dedicated source of revenue. Although housing trust funds may receive allocations of general revenue, corporate contributions, gifts, or other revenue, particularly to capitalize the fund at its onset, the key is that there is some source of revenue that is dedicated through legislation or ordinance. The dedicated revenue is usually a tax or fee, interest from accounts that are continuously established, or renewable resources. Although the funds may still need to be appropriated, it would nonetheless take an equivalent legislative action to undo the fund and its source of revenue. This distinguishes a housing trust fund from a revolving loan fund, annual appropriations for housing programs, or even a housing partnership that relies largely on corporate contributions. The appeal of housing trust funds to housing advocates and often government officials, as well as elected officials, is that it removes the constant search for housing funding from annual budget battles. If housing is a right for all citizens, or if it is seen as an essential ingredient in ensuring the health and safety of our residents in neighborhoods that provide affordable decent housing for all segments of the community, then it is appropriate to elevate funding for housing and give it more security. This is precisely what a housing trust fund does.

Approximately one fourth of existing housing trust funds have been able to dedicate more than one revenue source to the fund. This not only ensures increased revenues to support needed housing but also assists in leveling the impact that the larger economy has on any individual revenue source, such as a real estate transfer tax from which revenues may decline as markets decline.

2. A local program. City, county, and state housing trust funds are typically administered by an existing agency or department that has historically operated other housing programs, such as the federal Community Development Block Grant Program or the HOME Program. Housing trust funds do not rely on federal funding, however. They are local programs, using local funding and administered locally.

A few housing trust funds have gone to an entity outside government to administer the housing trust fund, such as a community foundation, that has the experience of administering distinct funds, reviewing applications, and monitoring projects. Some housing trust funds have incorporated their funds as distinct bodies with a governing board.

Nonetheless, in virtually all cases, administration of the fund is tied to local government either through the staff administering the fund, the board that oversees the operation of the fund, or both. This should be expected because the fund is dependent on a public source of revenue.

The costs of administering housing trust funds vary substantially in relation to the size of the trust fund and the level of technical assistance provided by the staff. Some housing trust fund legislation permits a portion of annual revenues dedicated to the fund to be used for administrative costs, typically 10% to 15%. Others support these costs through budgeted funds for staff administering programs, such as Community Development Block Grant funds or General Funds.

Most housing trust funds are overseen by an appointed board that acts in either an advisory capacity or actually makes decisions about the operation of the fund, including which awards are to be made. The boards are usually appointed by the mayor, city council, county commissioners, the governor, the legislature, a combination thereof, or in a few instances a department head. The boards are typically broadly representative of the housing industry with seats for financial institutions, real estate, private and nonprofit housing developers, housing service providers, housing advocates, the beneficiary population, and others.

3. Flexibility in program requirements. Most housing trust funds announce the availability of funds twice a year through a request for proposal process. The application specifies who is eligible to apply for funds, what kinds of projects are eligible, and what requirements the project must meet. Projects then compete for available funds and are often ranked based on characteristics of the proposal and priorities established by the trust fund.

Program guidelines are defined, broadly allowing nonprofit developers, for-profit developers, housing authorities, units of government, and often quasi-public entities to apply for funds to support projects that include new construction, rehabilitation, or conversion of structures that provide single or multifamily, rental or homeownership housing. These eligibility requirements tend to be inclusive. More restrictive are the specific requirements that the projects must meet to be eligible, in particular the population served.

Alternatively, less than one third of housing trust funds establish specific programs, allocate a portion of the available revenue to each program, and announce the availability of funds in various program rounds. For instance, a jurisdiction may want to establish a program to support the acquisition and renovation of vacant boarded-up homes and utilize a portion of the funds for this purpose.

Most housing trust funds provide loans or grants to sponsors of housing affordable to households earning no more than 80% of the median area income. A large proportion of trust funds either limit all their funds or a portion of them to a lower income, such as 50% or 30% of the median area income to ensure that these needs are also addressed. Housing trust funds are also attentive to other goals, such as maintaining the affordability of assisted units for a long period of time, leveraging housing trust funds dollars through other available funds, and providing needed housing-related services to support the targeted population.

4. Building partnerships. Particularly when the provision of housing for the poorest of households is the goal, housing trust funds rely heavily on the active participation of several players. On average, for every dollar from trust fund revenues supporting a project, one can expect $5 from other sources, such as financial institutions, federal home loan banks, federal low-income tax credits, other state or federal programs, foundations, and other sources. Thus, the coordination of housing trust fund revenue with other available housing dollars is important, and several trust funds do coordinate their programs with the federal HOME Program, the Community Development Block Grant Program, or both to maximize the use of all resources. Other housing trust funds work cooperatively with the Affordable Housing Program of the federal home loan bank system or with other local banks, corporations, or foundations.

Nonprofit housing development organizations are another important participant in housing trust funds. Providing housing for very low-income households and ensuring the long-term affordability of these units is a goal that most nonprofit organizations share with a housing trust fund. These organizations are eager to make use of available resources for needed projects. Although the overall capacity of nonprofit housing development organizations throughout the country has significantly increased (Keating, 1989; Peirce & Steinbach, 1987), this expertise does not exist in all jurisdictions or may be particularly lacking in selected areas of a city, county, or state, such as rural areas. To address the need for and interest in developing a capacity to build low-income housing, many housing trust funds have reserved some of their funds for capacity-building programs. These programs enable eligible organizations to add staff, obtain training, hire consultants, pay for predevelopment costs, or whatever has prevented them from entering the real estate development market. Developing the capacity of these organizations helps guarantee that the trust fund can achieve its goals.

Local Experiences in Creating Housing Trust Funds

The remainder of this chapter describes the development of housing trust funds in Chicago, Boulder, and San Diego. These three were selected, because they were created under different circumstances and demonstrate the adaptability of housing trust funds (see Table 10.2).

Chicago's Low-Income Housing Trust Fund: Housing Activists Fight for Housing for Their Poorest

Chicago's housing trust fund was created through the efforts of several coalitions of housing and homeless advocates. After years of struggling unsuccessfully to secure agreement on a possible revenue source for such

Table 10.2. A Comparison of Three Housing Trust Funds

	Chicago, Illinois	*Boulder, Colorado*	*San Diego, California*
Population	7.5 million +	80,000	1 million +
City median income	$30,707	$41,700	$43,900
Poverty rate	19.0%	6.9%	6.9%
Date trust fund enacted	1989	1991	1990
Development of fund initiated by	Housing activists	City Task Force	City staff and housing activists
Average annual expenditure of trust fund	$700,000	$1 million	$6 million
Administration of fund	Department of Housing staff board of trustees	Department of Housing and Human Services staff advisory board	Housing Commission staff board of trustees
Uses of the fund	Rental subsidies to projects	Loan or grants for projects allowing for wide latitude in project development with priorities for assisting families with children, acquisition of existing housing, transition to nonassisted housing, and housing for large families	Loans or grants to projects for rehabilitation, housing development, mobile homes, first time homebuyers, rental assistance, transitional housing, and nonprofit capacity building
Intended beneficiaries	Households at or below 50% of median-income	Households earning between 30% to 60% of median-income	Households at or below 80% of median-income
Revenue sources	Interest from bond fees used to capitalize the fund. Fees charged for first time homebuyers program	8/10th of a mill increase in property tax. Excise tax on new residential and nonresidential development	Impact fee on nonresidential development projects. Redevelopment tax increment funds appropriated in lieu of transient occupancy tax originally committed to fund
Total units supported	300+ units assisted annually	150 units completed, 120 units planned	1283 units, 813 beds, 95 mobile homes
Average dollar leveraged per project for each dollar committed from trust fund	not applicable	$5 for every $1 from the housing trust fund	$5 for every $1 from the housing trust fund

a fund, the focus turned to Presidential Towers, a private development of 880 new luxury apartments that resulted in the demolition of the last remaining single-room occupancy hotel on the near west side. Presidential Towers symbolized all that was wrong with development priorities in Chicago, according to housing advocates: displacement of more than 2,300 low-income people; development of upper-income housing with public subsidies going to highly profitable private real estate ventures; and an emphasis on the Loop rather than on neighborhoods (Weiner, 1987). In 1987, after months of activities by housing advocates, the city reached an agreement with Presidential Towers that in exchange for city financing at lower interest rates, the developers would place funds into a low-income housing trust fund.

The agreement was to result in $14 million to $17 million committed to the fund over several years in return for the city issuing tax-exempt bonds to finance a future expansion, as well as the refinancing of the initial phase of Presidential Towers. The city's coalitions[1] succeeded in incorporating into the ordinance assurances that the fund would be targeted to residents with incomes at or below 50% of the city's median income, and that community groups and low-income residents would be involved in establishing and implementing the trust fund. In addition, it was believed that for the trust fund to be truly effective, there needed to be an ongoing resource committed to it. Indeed, the ordinance states, "It shall be the goal of this corporation to expand the capital base of the corporation to between $20-25 million within five years after the corporation is legally incorporated."[2]

In reality, Chicago's Low-Income Housing Trust Fund received only $3.2 million from Presidential Towers bond fees, primarily because of financial difficulties that the project encountered, and has been operating off the interest of this initial investment. The fund also received $1 million from a special Congressional earmark and $2.6 million in federal HOME funds for fiscal years 1992 and 1993. In addition, a portion of the fee charged to first-time homebuyers participating in a new Chicago program goes into the trust fund. The first-time homebuyers program provides direct tax credits of 35% of mortgage interest up to $2,000 per year to qualified homebuyers. Of the fee required of homebuyers to offset the initial administrative costs associated with the program, equal to 1.75% of the mortgage financing, one fourth goes into the trust fund (City of Chicago, 1991, 1992).

The Fund is intended to supplement other existing sources of support for housing and to expand the pool of housing units available to very low-income residents. Funding is available to both nonprofit and for-profit developers who create and maintain decent and affordable housing. The funding can take the form of grants, loans, or other types of financial assistance for individual projects. Types of projects that are eligible for funding include but are not limited to operating cost subsidies, rent subsidies, mortgage loans, and the acquisition or rehabilitation of single and multifamily existing housing or both. However, to date, the fund's priorities have been limited to allocating funds for rental subsidies. Beginning in 1992, the fund launched a new program providing cash grants or low-interest loans to developers of affordable housing to help offset the long-term cost of operating rental apartment buildings. The one-time, larger infusion of funds is to allow building owners to lower rents for a portion of their units over a long-term period. Previously, developers and nonprofit organizations were granted funding by the Trust Fund on a per-year basis, requiring developers to reapply for funding each year.

Each year, Chicago's Low-Income Housing Trust Fund has committed around $700,000 or more to provide rental subsidies to more than 300 units of affordable housing. The monthly per unit allocations ranged from $94 to $443 with the average monthly subsidy just under $200 in 1991. Although the Trust Fund is incorporated with a board that oversees its operations, it is staffed by the city's Department of Housing.

In 1993, housing advocates in Chicago succeeded in their Housing and Community Jobs Campaign, winning a commitment of increased spending for affordable housing by 50% over the next five years to $750 million. Included is a commitment for $20 million over the next five years for the Low-Income Housing Trust Fund (Chicago Rehab Network, n.d.).

■ Boulder Community Housing Assistance Program: Planning for Assisted Housing in a Growth-Controlled Community

In 1989, the City of Boulder created an Affordable Housing Task Force to study and recommend solutions to community housing needs. It was to evaluate an existing housing program, based on an inclusionary zoning ordinance, and make recommendations to the city council that would make housing more affordable in the city. After ten months of work, the Task Force made five recommendations, including one that resulted in the creation of the city's housing trust fund based on residential and commercial development taxes and an increase in property taxes.

The Boulder development community was concerned about affordable housing, and there was widespread recognition that a community-wide answer to creating a stable workforce through a diverse housing supply was needed. Boulder is a growth-controlled community and the Task Force concluded that enabling Boulder to establish an assisted housing pool that would equal 5% of the total housing stock was a reasonable goal.

It proposed a housing trust fund that would produce 46 assisted housing units and rehabilitate 17 additional units per year. A per unit subsidy cost of $20,625 was proposed, requiring a $1.3 million annual fund. This would enable the city to reach its 5% goal within 25 years. The Task Force, supported by city staff, collected extensive data and held meetings with a variety of groups and organizations. A local survey revealed that 57% of the sample surveyed supported local government's role in making housing more affordable.[3]

City Council went along with most of the recommendations from the Task Force. Although changing the character of the program by making its target population 30% to 60% of the city's median income, they implemented the program by creating an excise tax on residential and nonresidential development and increasing the local property tax by 8/10th of a mill.

Although the funds are not dedicated by ordinance, which is typical of most housing trust funds, the program is too popular to be withdrawn by City Council. The Community Housing Assistance Program provides the first dollar in for a proposed project and, as a consequence, the program will fund "on concept." Although most proposals are well developed, the sponsor does not even need to have land to be considered. Funds are available to nonprofit organizations, private developers, churches, individuals, the local housing authority, neighborhood based groups, and public-private partnerships and can take the form of grants or loans, deferred loans or project equity, depending upon the nature of the project. The program has exceeded its goal and has supported 150 units in three years with another 120 units in various planning stages.[4] Boulder's Community Housing Assistance Program receives approximately $1 million in revenue each year.

Program priorities are established every two years in coordination with the adoption of the two-year work plan. Current priorities include assisting families with children, particularly those earning 30% to 40% of the area median income; acquisition of existing housing; encouraging transition to nonassisted housing; and providing opportunities for housing large families. No more than 35 assisted units can be located within one development and the program encourages mixed-income projects. Funds are released through a request for proposal process and is now coordinated with other funds administered by the Department of Housing and Human Services, including the Community Development Block Grant Program and the federal HOME program. Applicants may request funds from all three funding sources for one project.

San Diego Housing Trust Fund: Planners and Activists Cooperate to Beat the Odds

The San Diego Housing Trust Fund was created through the joint efforts of staff from the city's Housing Commission and the advocacy of community groups. Although numerous community groups, nonprofit housing development organizations, fair housing groups, and other housing advocacy organizations existed in San Diego, it was the staff of the Housing Commission that first considered the notion of a trust fund. In 1988, the Commission brought more than 30 community groups together to educate them about the potential of a trust fund and to engage their support (Calavita & Grimes, 1992, p. 176).

Subsequently, community groups organized the San Diego Housing Trust Fund Coalition and identified four points that they wanted to become a part of the housing trust fund: It should help house those most in need, the community should have control over the fund, the fund should be used to support nonprofit development, and continued affordability of the units assisted should be ensured. Each made its way into the final trust fund, adopted two years later (Calavita & Grimes, 1992, p. 176).

In the spring of 1989, the Housing Commission approved in concept the establishment of a housing trust fund and established a task force to make recommendations as to how such a fund could be developed. Several factors contributed to the eventual creation of the fund. A high-growth period, coupled with inadequate infrastructure and public facilities, created an environment highly susceptible to such a fund; a district elections initiative brought about a new City Council; the Task Force chair became a strong advocate of the fund and as a developer brought about consensus on numerous issues; and the Task Force meetings were open to the public and attended almost exclusively by Housing Trust Fund Coalition members, bringing prepared positions on whatever was being discussed and never letting Task Force members forget that there was a housing crisis for poor residents of the city (Calavita, Grimes, & Reynolds, 1994).

The Task Force hired a consultant to provide recommendations on revenue sources, governance, and beneficiaries of the proposed fund. Research on revenue options resulted in the recommendation of a package of five sources that called on all segments of the community: a citywide commercial development fee; a portion of the annual increment of the hotel room tax; a gross receipts business tax; a landscape, lighting, and park maintenance fee; and a 2% utility users' fee. In its final report, the Task Force recommended a trust fund of $54 million, estimated to be necessary to meet one third of the housing need by the year 2000.[5]

Under severe budget constraints and sizeable opposition, the City Council reduced the size of the fund to $13 million collected from only two of the proposed sources: the commercial development fee and a portion of the hotel room tax. Nonetheless, in 1990, San Diego passed its housing trust fund, administered by the Housing Commission with an 11-member Board of Trustees appointed by business, industry, labor, and nonprofit and community-based organizations.

The distribution of funds is established annually through a Three-Year Program Plan that divides available resources among rehabilitation, housing development, first-time homebuyers, mobile home parks, transitional housing, and nonprofit capacity building efforts. In 1993, more than $7 million was awarded to 30 projects.[6]

Despite its success in addressing a wide range of housing needs throughout the city, San Diego's Housing Trust Fund has never freed itself of controversy. In the summer of 1992, the City Council diverted the hotel room tax revenues to help solve the city's fiscal woes. In return, the Council allocated $3.2 million from its Center City Development Corporation (San Diego's Redevelopment Agency) tax increment housing set aside to the trust fund. In 1993, although housing advocates sought to restore the hotel room tax to the trust fund, the Council voted again to provide tax increment funding and merely left intact language allowing the hotel room tax revenues to be used for the Trust Fund. Funding for the Trust Fund is now restored to $5 million to $6 million in annual revenue.[7]

Conclusions

It is important that housing trust funds are designed with a few key parameters in mind—such as targeting funds to the lowest incomes, citizen involvement in the administration, and objectives of long-term affordability for the assisted units. Nonetheless, housing trust funds tend to be adaptable. They fit local situations, because they are designed locally and use local resources, unlike most assisted housing programs that have been designed at the federal level and funded with numerous restrictions attached.

Housing trust funds have been able to develop strong partnerships with local nonprofit housing development organizations, to leverage other available housing resources, and to address difficult housing needs requiring creative approaches. They range from committing more than $100 million to less than $100,000 annually. The total amount of funds they have to disburse each year is far less than what is needed. Although housing trust funds nationwide each year commit hundreds of millions of dollars, most are relatively small. And they are relatively new. More than half of the housing trust funds currently in operation have been created since 1990.

Nonetheless, their impacts are noteworthy. Each fund can document hundreds of units assisted. Funds have been creative in the design of their programs and have supported innovative projects, serving special populations, extremely low-income households, and building new capacities. They have developed procedures that enable funds to be accessed easily and quickly. They have brought other public and private dollars into the housing arena and they have had a positive impact on local economies.

On average, a housing trust fund takes a minimum of two years to develop. The three case studies demonstrate how much effort may be required. Chicago's was created through the initiative of citizens, Boulder's through the city's concern, and San Diego's through a combination of both. None came easily. Each required the development of a strong proposal that was well researched, that outlined the details of the proposed program, and that developed clear arguments for the need for a housing trust fund. Each required the creation of political will to vote the fund into existence. In Chicago, this took pressure from the voters on their elected officials. In Boulder, it took the development of a proposal that had both popular and political support. In San Diego, it took all of this.

Virtually every housing trust fund proposal faces opposition. It usually comes from whatever industry is most directly affected by the proposed revenue source. A good proposal with considerable popular backing can lose, because the opposition is more powerful and more influential with elected officials. Most housing trust funds require a well-thought-out and well-executed campaign that convinces elected officials that the need for a housing trust fund overrides the opposition, and that there is sufficient support to make their positive vote a palatable one.

The experience of Chicago, Boulder, and San Diego further demonstrates that the creation of a housing trust fund is complex. Its design requires detailing an administrative process that is efficient, programs that address the existing needs, procedures that protect key objectives of the fund and ensure successful projects, and revenue sources that are accessible, stable, and productive. Its creation requires a campaign that includes developing concise information about the proposal, coalition-building involving a wide range of housing interests, negotiation over the final details of the ordinance or legislation, lobbying to get the votes needed for passage, working with the media to ensure accurate public information, dealing with the opposition, and staging other public events to draw attention to the issue.

As Chicago, Boulder, and San Diego demonstrate, housing trust funds are political. The single most decisive factor in getting a housing trust fund passed is the degree of political will behind it. Whether this comes from the citizenry or the elected officials, without it, housing trust funds are rarely created.

NOTES

1. Statewide Housing Action Coalition, the Coalition for the Homeless, the Balanced Growth Coalition, and the Chicago 1992 Committee.

2. An ordinance authorizing the establishment of a *Chicago Low-Income Housing Trust Fund* and approved Articles of Incorporation and By-Laws for that not-for-profit corporation, passed June 28, 1989. The Trust Fund was incorporated as a nonprofit organization on February 27, 1990.

3. *Report to Boulder City Council on Affordable Housing From the Affordable Housing Task Force and the Department of Housing and Human Services,* February 13, 1990.

4. Kathy McCormick, City of Boulder, Department of Housing and Human Services, interview, July 27, 1994.

5. *Creating Affordable Housing for San Diegans, Final Report and Recommendations,* San Diego Housing Trust Task Force, September, 1989.

6. San Diego Housing Commission, *Second-Year Activities Report: San Diego Housing Trust Fund,* Spring, 1993.

7. *San Diego Advocates Win Fight to Keep Funds, News From the Housing Trust Fund Project* (Washington, DC: The Center for Community Change, July, 1994), pp. 4-5.

REFERENCES

Brooks, M. E. (1989). *A survey of housing trust funds.* Washington, DC: Center for Community Change.

Brooks, M. E. (1994). *A summary of revenue sources committed to existing housing trust funds.* Washington, DC: Center for Community Change.

Calavita, N., & Grimes, K. (1992). The establishment of the San Diego Housing Trust Fund: Lessons for theory and practice. *Journal of Planning Education and Research, 11*(3), 170-184.

Calavita, N., Grimes, K., & Reynolds, S. (1994). Zigzagging toward long-term affordability in the Sun Belt: The San Diego Housing Trust Fund. In J. E. Davis (Ed.), *The affordable city: Toward a third sector housing policy* (pp. 265-291). Philadelphia, PA: Temple University Press.

Chicago Rehab Network. (n.d.). *Chicago affordable housing and community jobs campaign '93 victory!* Chicago Rehab Network.

City of Chicago, Department of Housing, Chicago Low-Income Housing Trust Fund. (1991). *Providing new beginnings: 1991 annual report.* Chicago: Author.

City of Chicago, Department of Housing, Chicago Low-Income Housing Trust Fund. (1992). *Providing new beginnings: 1992 annual report.* Chicago: Author.

Fabricius, M. A., & Snell, R. K. (1990, September). *Earmarking state taxes.* Denver, CO: National Conference of State Legislatures.

Hebert, S., Heintz, K., Baron, C., Kay, N., & Wallace, J. E. (1993, November). *Nonprofit housing: Costs and funding final reports.* Washington, DC: U.S. Department of Housing and Urban Development.

Keating, W. D. (1989, February, March, April). The emergence of community development corporations: Their impact on housing and neighborhoods. *Shelterforce,* pp. 8-14.

Leonard, P. A., Dolbeare, C. N., & Lazere, E. B. (1989). *A place to call home: The crisis in housing for the poor.* Washington, DC: Center on Budget and Policy Priorities and the Low-Income Housing Information Service.

Peirce, N. R., & Steinbach, C. F. (1987). *Corrective capitalism: The rise of America's community development corporations.* New York: The Ford Foundation.

Weiner, D. (1987, January/February). Presidential Towers agreement reached. *The Network Builder.* Chicago, IL: Chicago Rehab.

11 Learning From Experience: The Ingredients and Transferability of Success

WILLEM VAN VLIET– –

Abstract

This chapter pulls together those that precede it. A brief review of current housing problems in the United States serves as a backdrop for a sketch of the broader context of affordable housing provision and urban redevelopment in large cities. This context is significantly shaped by global restructuring and national policy shifts, which have reinforced patterns of socioeconomic polarization and spatial segregation. The case studies in this book illustrate how local communities can mitigate those effects. The chapter concludes with a discussion of how other communities can draw lessons from these experiences.

■

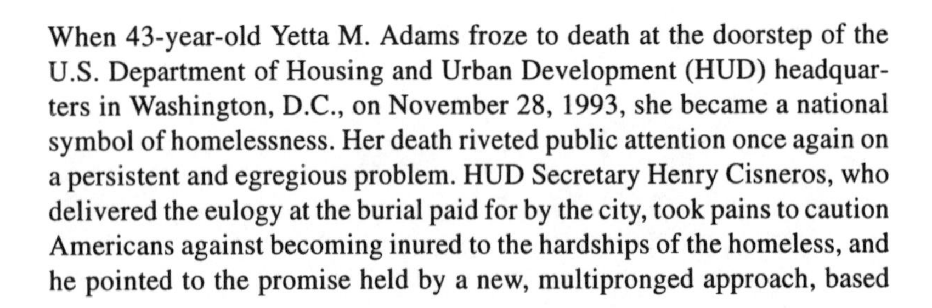
When 43-year-old Yetta M. Adams froze to death at the doorstep of the U.S. Department of Housing and Urban Development (HUD) headquarters in Washington, D.C., on November 28, 1993, she became a national symbol of homelessness. Her death riveted public attention once again on a persistent and egregious problem. HUD Secretary Henry Cisneros, who delivered the eulogy at the burial paid for by the city, took pains to caution Americans against becoming inured to the hardships of the homeless, and he pointed to the promise held by a new, multipronged approach, based

AUTHOR'S NOTE: Rachel Bratt, Susan Clarke, Dennis Judd, Peter Marcuse, Mary Nenno, and Mara Sidney read an earlier version. I am grateful to them for helpful comments.

on emerging forms of innovative intergovernmental cooperation and public-private partnerships (Cisneros, 1993).

The preceding chapters detail the formation of such coalitions, the mobilization of resources, and other aspects of the difficult and often long struggles to make severely distressed neighborhoods into viable communities, to turn failure into success. This last chapter pulls together these diverse cases, placing them in the context of broader economic and political developments. First, it briefly reviews the nature and distribution of current housing problems in the United States. These problems are then interpreted against the background of global restructuring and shifting national policy orientations, trends that magnify polarization and imply greater responsibilities for cities. The chapter examines local communities' responses to these recent challenges and identifies elements that contribute to success. A discussion of the transferability of models that appear to work concludes the chapter.

Housing Problems

In attempts to solve housing problems, it is important to recognize success stories. Indeed, the lessons they hold form a central theme of this book. Yet informed decision making must also be based on a recognition of how policies fall short. And, by any reasonable standard, the nation's policies have failed to eliminate a number of persistent housing problems. This is readily illustrated by even a cursory review (for more detailed recent reviews, see Burchell & Listokin, 1995; Dreier & Atlas, 1996).

Homelessness

Estimates of the number of homeless people vary according to how, when, and where people are counted (Marcuse, 1990; see also Burt, 1995; Hopper, 1995; Wright & Devine, 1995). Whatever the method used, studies have typically relied on information collected at one particular point in time. The resulting estimates overrepresent the incidence of long-term homelessness, but underrepresent the incidence of homelessness overall.[1] A recent nationwide survey, avoiding biases built into the methods of these studies, indicates that homelessness is much more widespread than commonly thought (Link, Phelan, & Bresnahan, 1994). Findings from this research suggest that 14% of all Americans (26 million) have experienced homelessness at some point in their lives and that 4.6% (8.5 million) were homeless at least once during the 1985-1990 period.

Rental Burden

Aside from those who are homeless, many more households experience serious affordability problems. In 1991, 33% of very-low-income renters paid more than 50% of their income on housing; 16% lived in structurally inadequate housing. Among extremely low-income renters, these percentages were higher yet: 70% and 22%, respectively (Joint Center for Housing Studies, 1994). Figures provided by HUD show that from 1978 to 1991 there was a 50% increase in households paying in excess of one half of their income on rent or living in substandard housing (HUD, 1995c). In 1993, among renters living in poverty, 35% of Hispanics and 51% of Blacks paid 50% or more of their income on rent (HUD, 1995c); 25% of all renters lived in poverty (36% among Hispanics and 42% among Blacks).

At the same time that the number of poor renters has increased, the size of the low-cost rental stock has decreased. From 1986 to 1991, there was an annual net loss of 130,000 low-cost rental units (Joint Center for Housing Studies, 1994, p. 16). Yet current housing and income transfer programs serve only a fraction of those eligible for assistance. In 1991, just over one third of extremely low-income renters received some form of housing subsidy, as did only one fifth of very-low-income renters and a mere 6.5% of low-income renters.[2]

Barriers to Home Ownership

For those seeking to own a home, the trend is no more encouraging. Data compiled by Harvard's Joint Center for Housing Studies (1994) reveal a continuing decline in the rate of home ownership among all groups under the age of 55, except most elderly. The decreases are especially strong in the 25-to-35 age group. Young households make up a growing proportion of the nation's poor. Their incomes, in constant dollars, are now less than they were 20 years ago. Adaptive regulatory innovations (e.g., counting a second income in loan applications; the introduction of alternative mortgage instruments, used for 36% of all loans in 1995) have not reversed this downslide in private home ownership rates. Lower interest rates have recently improved affordability, but the down payment and closing costs, amounting to 59% of first-time buyer income in 1993, keep ownership beyond the reach of many. The median income of first-time buyers as a proportion of that of repeat buyers declined from 98% in 1976 to 83% in 1984 to 73% in 1994. Their average age has risen (from 28 in 1976 to 32 in 1995), and they need more time to save for the down payment, even though a decreasing portion of it comes from their own savings (1976: 80%; 1995: 75%). The average monthly payment of first-time buyers, as a percentage of after-tax income, has increased from 23% in 1976 to 30% in 1984 to 35% in 1995 (Chicago Title and Trust Company, 1996).

Discrimination

Affordability problems of renters, as well as owners, are made worse by institutional and market-based discrimination. Racial steering remains widespread, although it has been illegal since passage of Title VIII of the 1968 Civil Rights Act. Recent fair housing audits have found evidence of racial steering of between 20% and 60% of rental and sales transactions with real estate agents (Yinger, 1995).

As amended in 1976, the Federal Equal Credit Opportunity Act makes it unlawful for a lender to discriminate on the basis of the race, color, religion, national origin, sex, marital status, or age of an applicant in any aspect of a credit transaction. However, mortgage loans remain less available to some than to others. According to data released by the Federal Financial Institutions Examination Council in 1993, banks, savings institutions, and credit unions turned down 34% of the mortgage applications received from Blacks, 28% from American Indians, 25% from Hispanics, and 15% from Whites. The disparity in rejection rates remained wide after adjusting the data for differences in income (see also Myers & Chan, 1995).

Although research suggests gradual improvement (Ladd, 1982), women also face barriers when applying for mortgage loans. According to the Office of the Comptroller of the Currency, banks can discriminate by refusing to count alimony or child-support payments as income, or by requesting that a woman have a cosigner even when she has enough income to qualify for a loan (O'Connell, 1994; cf. Taylor & Jureidini, 1994). Women who are victims of domestic abuse have also been denied mortgage insurance on the grounds that coverage would be an incentive for batterers.[3] Furthermore, restrictive rental practices, justified as rational risk-aversive business decisions, and exclusionary zoning make a large portion of the rental stock inaccessible to families with children (Marans & Colten, 1985; Ritzdorf, 1986).[4] The bottom line is that all of these discriminatory practices reinforce patterns of spatial segregation and socioeconomic polarization, creating unequal life chances for minorities, women, low-income people, and young families.

The case studies in this book and other research (e.g., Kasarda, 1993; Vidal, 1995) make clear that the distribution of these and other housing problems is neither random nor even. Rather, they are often concentrated in areas beset by other problems as well. We frequently find them in racially segregated, declining, central-city neighborhoods with inadequate municipal services; selective out-migration; rampant abandonment; a defective physical infrastructure; an insufficient tax base; a scarcity of community institutions; high rates of poverty, unemployment, and crime; lack of medical care; and unsatisfactory access to education. Understanding the interconnections among these multiple problems, as well as their broader context, is essential to efforts aimed at resolving them.

The Broader Context of Housing Problems

In and of itself, "affordable housing" means very little. The cost of housing becomes affordable only in relation to the income of its occupants (e.g., Stone, 1993). Although the absolute housing shortage in the United States has long been eliminated, relative deficits persist, in part caused by mismatches between housing and job markets (Ihlanfeldt, 1994). Therefore, we must think of housing in connection with the availability and types of employment opportunities. Affordable housing cannot be produced without consideration of the broader contexts affecting the price of housing and the earning power of households.

Which factors, then, set the context for housing? The discussion in this section will focus on two levels. First, it will briefly review developments at the international level, in particular the trend of global restructuring. Second, it will highlight national policy shifts, in particular the trends of decentralization and privatization. Finally, this section will conclude that these changes imply greater socioeconomic polarization and spatial isolation, resulting in growing concentrations of population groups that are marginal to society's economic, political, and social processes, raising the question of how cities can mitigate these local effects.

Global Restructuring

Today's urban problems do not occur in a vacuum. They are significantly framed by the restructuring of the global economy. This process involves new, more flexible patterns of production, facilitated by technological innovations, improved informational capabilities, and increased capital mobility. An extensive, multidisciplinary literature exists on these aspects of restructuring. Here, however, the concern is with the implications of global restructuring for the composition and distribution of urban populations.

In this connection, a basic thesis, notably advanced by Sassen (e.g., 1991, 1994), states that socioeconomic polarization and spatial segregation in global cities have increased as a result of the changing structure of employment. Manufacturing jobs with solid pay have given way to a bifurcated service sector with low-paying, unskilled, dead-end, service jobs and high-paying, specialized, technical and administrative jobs.[5] It has also been argued that advanced communication techniques (telematics) reduce the costs of financial transactions and increase the sensitivity of markets across national borders, making existing governance structures obsolete and undermining the capacity of the state to produce public goods (Cerny, 1994). Against this background, the local provision of affordable housing has often been dominated by massive, largely international investment in commercial real estate development, which has further sharpened patterns of inequality and the concentration of poverty (e.g., Fainstein, 1994; Harloe, Marcuse, & Smith, 1992). In combination, these processes have marginalized economically and politically vulnerable population groups, sometimes referred to as the underclass (e.g., Wilson, 1987).[6]

What is of interest in the context of this book is the potential of public policies in general, and housing policies in particular, to mitigate (or, conversely, to accentuate) the polarizing tendencies of global restructuring (cf. Heisler, 1994; Lee, 1994).

National Policy Shifts

Only 20% of Americans "trust the government in Washington to do what is right" most of the time, down from 73% in 1958 (Ladd, 1995). The federal government, especially under President Reagan, seized on this national mood to advance the "new federalism." Under this banner, it sought to divest itself from important responsibilities in areas where it had previously played leading roles. It seemed politically expedient and economically prudent to devolve federal functions to the state and local levels, often in the form of unfunded mandates or by just simply cutting aid. As a result, from 1980 through 1992, the balance of intergovernmental aid to cities, after excluding the share of states (fluctuating around 16.5%), declined by 68.5% to 5.4% of total revenue (U.S. Bureau of the Census, 1995, p. 314). This trend continues into the present under a Congress dominated by Republicans.

The implications for cities of the recent proposals are not only less funding but also greater inequality. The reliance on block grants may erode the need-based targeting designed into present federal standards. The effect will likely be greater disparities among states.[7] Because decentralization typically gives local officials greater autonomy over fewer resources, their actions also become more susceptible to the vicissitudes of market processes, thus creating greater potential for local conflict (Clarke, 1989). Decentralization appears to reduce the likelihood of interlocal cooperation as well (Wrightson, 1986).

Various definitions of privatization exist (e.g., Van Vliet--, 1990). For the present purpose, one can describe this process as a shifting of public responsibilities to the private sector, which may include the for-profit as well as the nonprofit sector. An example is HUD's demonstration program to sell off a portion of the public housing stock, an experiment that was doomed largely because the sitting tenants were too poor to buy their units (Rohe & Stegman, 1990). Under the Reagan administration, privatization went hand-in-hand with draconian budget cuts. From 1980 through 1988, budgetary authority for federal low-income housing assistance programs was slashed by 80%. But housing was not the only target of the Republican cutbacks. Between 1979 and 1990, social welfare expenditures under Title XX were nearly cut in half (Stoesz & Karger, 1993), cash benefit programs (food stamps, AFDC) were curtailed, and eligibility for public entitlement programs was restricted. During this period, the number of persons living in poverty grew from 29 million to 39 million (U.S. General Accounting Office [GAO], 1995). The underlying ideology was succinctly captured by David Stockman, former President Reagan's first budget director and a major protagonist of "free" markets. Opposing entitlement to any kind of service, he stated: "I don't accept that equality is a moral principle" (quoted in Kozol, 1988, p. 163). Since then, further belt-tightening has occurred.

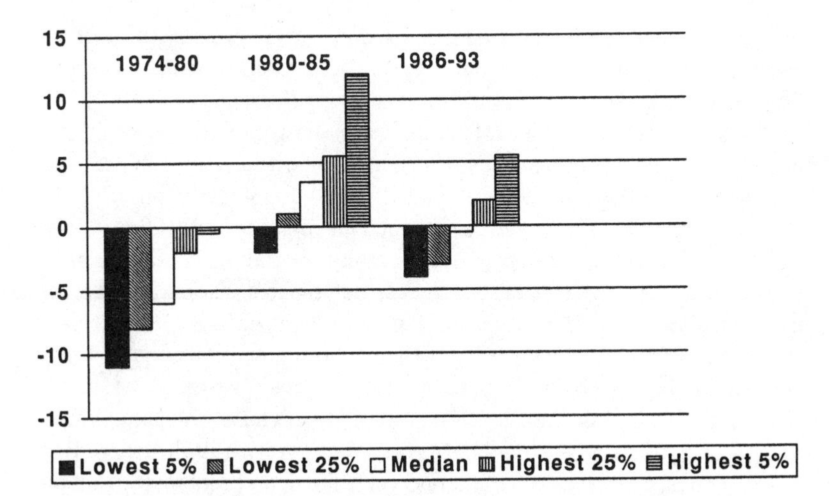

Figure 11.1. Household Income Change (%)
SOURCE: Joint Center for Housing Studies (1994).

Socioeconomic Polarization and Spatial Isolation

Global restructuring of the economy has increased the potential for greater socioeconomic polarization and spatial isolation at the local level. However, this is no simple cause-and-effect relationship. The specific outcomes in any given city are not predetermined but, rather, modulated by national and local policies (Burgers & Kalb, 1995; Goetz & Clarke, 1993). As just noted, in the United States, federal policies, and the lack thereof, have done little to cushion the effects of global restructuring on vulnerable population groups. Instead, the shifts toward decentralization and privatization have contributed to a worsening of existing problems.

Numerous statistics attest to increasing disparities among population groups. For example, during the 1980s, federal tax rates increased by more than 25% for households in the bottom income quintile, whereas they decreased by nearly 10% for the top quintile. Figure 11.1 shows household income change from 1974 to 1993. Throughout this period, median income of the poorest 5% of the population continued to fall. The fastest growth, on the other hand, is recorded for the highest income groups. A recent OECD study found the middle class in the United States smaller by far than that in 13 other industrialized countries, whereas its income inequality is greatest (Henwood, 1996). The bottom stratum comprises disproportionate numbers of non-White and young households (Wolff, 1995).

Housing tenure lines also reflect income polarization. In 1970, the median income of renters was 64.9% of that of owners; in 1980, it had declined to 53.0%; in 1990, it had dropped to 50.3%, and in 1993 even further to 47.8% (Joint Center for Housing Studies, 1994, p. 24). Income polarization similarly occurs along spatial lines. Of all U.S. households, 31% live in central cities, but the comparable figure for poor households is 42%, and the gap is growing (HUD, 1995a). One recent study found that between 1970 and 1990 segregation of the poor increased by about 10%, as measured by isolation and dissimilarity indexes (Abramson, Tobin, & Vander Goot, 1995). Others have obtained similar results (e.g., Jargowsky, 1994; Massey & Eggers, 1993). Examining poverty rates, as well as other distress variables such as joblessness, welfare receipts, and the teenage school dropout rate, Kasarda (1993) also reports patterns of increased spatial concentration of problems in the nation's 100 largest cities (see Table 11.1). Discriminatory housing practices reinforce the spatial isolation of the poor (Massey, Gross, & Shibuya, 1994).

In the absence of strong federal redistributive policies, the question arises: What about local policy? Can city governments act as a countervailing force? Not according to conventional arguments, as presented by Peterson (1981) in his often cited and aptly titled *City Limits.* As Savitch (1995) writes, "The size, shape, economic structure and social fabric of delocalized cities depend on what happens in the outer world" (p. 137). However, others have begun to advance less deterministic perspectives (see, e.g., Clarke, 1993; Logan & Swanstrom, 1990; Wilson, 1995). Indeed, recent literature provides empirical evidence that cities, in fact, can (and often do) play an important role in mitigating the polarizing and exclusionary effects of higher-level processes (e.g., Davis, 1994; Goetz, 1993).[8] The studies in this book further confirm that local policies can make a difference. So, naturally, one wonders how this happens. What accounts for success in cases where marginalized communities become reintegrated into the urban fabric? The preceding chapters describe a variety of approaches. Regardless of their differences in detail, these approaches tend to share a number of characteristics. Before reviewing these recurring features, we must consider the definition of success.

TABLE 11.1 Census Tracts by Poverty and Distress Status for the 100 Largest Cities, 1970-1990 (% of city total)

Census Tracts	*1970*	*1980*	*1990*
Poverty tracts[a]	27.3	34.2	39.4
Extreme poverty tracts[b]	6.0	9.7	13.7
Distressed tracts[c]	2.4	11.0	13.0
Severely distressed tracts[d]	1.3	4.1	4.0

SOURCE: Population Census Data for 1970, 1980, and 1990 reported in Kasarda (1993).
a. At least 20% of population below poverty level.
b. At least 40% of population below poverty level.
c. Population with disproportionately high levels of poverty, joblessness, female-headed families, and welfare receipts.
d. Same as distressed tracts, with, in addition, exceptionally high teenage school dropout rates.

The Definition of Success

From 1970 to 1990, the proportion of the U.S. population living in neighborhoods of extreme poverty increased from 5% to 11%. By its own admission, HUD produced "failed policies [that] contributed to a concentration of poor families in inner-city neighborhoods" (HUD, 1995c). However, the federally assisted successful redevelopment of some of those neighborhoods, detailed in this book, provides evidence that militates against such a blanket self-indictment. This raises the question, What constitutes success?

In a recent prizewinning study, Wolman, Ford, and Hill (1994) identified urban success stories based on the degree of agreement among the perceptions of a panel of experts regarding the strongest economic turnaround or urban revitalization among the 50 most distressed cities during the 1980-1990 decade.[9] They then compared the experts' ranking to the ranking produced by indicators of resident economic well-being in each of the cities.[10] Results of their analysis indicated that in the cities perceived by the experts as having been most successful, taken as a group, the economic well-being of the residents increased no more—and in many cases much less—than in other cities that were equally distressed in 1980. The data in this study do not permit a testing of the possible explanations that could account for this discrepancy between the subjective opinions of the experts and the objective reality of resident economic well-being. Nevertheless, one may speculate that the experts' perceptions were biased by their familiarity with a more noticeable form of urban revitalization, namely, physical redevelopment of the downtown. Yet improvements in resident well-being do not necessarily parallel the more conspicuous revitalization of the central business district. Indeed, in Chapter 8, Goetz offers evidence of how the latter may occur at the expense of the former (see also Frieden & Sagalyn, 1989).

How, then, does one define success? A vexing question. The authors in this book hesitate to articulate explicit criteria. This apparent reluctance reflects the difficulty of identifying agreed-on criteria of success. This difficulty derives, in part, from the severity and interrelatedness of problems in these communities. The chapters detail horrendous conditions. For example,

- during 1980-1990, the Lake Parc Place neighborhood lost 50% of its population, vacancies multiplied sixfold to one third of all units, two thirds of the population was below the poverty level, 60% was on welfare, and the median income ranked last among Chicago's 77 community areas;
- from 1970 to 1985, Baltimore lost 45% of its manufacturing jobs; the city experienced an exodus of White households, while Black median income as a percentage of the metro area median income fell to 55% in 1990; in Sandtown-Winchester, the infant mortality rate was double that of the city, 90% of births were to single mothers, and 50% of households were below the poverty level. Furthermore, only 10% of its elementary school students meet statewide performance standards, and 75% of the residents live in substandard housing (Wexler, 1995);
- between 1970 and 1990, West Farms, the Bronx, lost 60% of its population, which also underwent a radical change in racial composition; half lived in poverty; one quarter reported incomes under $5,000 per year; only 43% had graduated from high school; two thirds of households with children were headed by women; and hundreds of families were living doubled up;
- before its renovation, Commonwealth in Boston had a vacancy rate of 52%, 94% of its households were headed by women, and only 14% of the adult population was employed.

The preceding chapters portray conditions of abject poverty, widespread unemployment, rampant abandonment and disinvestment, pervasive overcrowding, rodent infestation, crime, and other threats to health and safety. In other words, this book documents situations, by no means unique, that are very bad, in more than one way. Against this background, success seems to be *any* change for the better, whether it be an increase in decent and affordable housing, the creation of a new funding source, a reduction in crime, the provision of local jobs, or improved municipal services.

However, on closer scrutiny, the outcomes can be more complex, depending on how the issues are defined. One must ask, for example, success for whom? Who is left out, perhaps worse off? Careful tenant selection played an important role in the success of Lake Parc Place in Chicago and Commonwealth in Boston. Other chapters similarly stress the importance of tenant screening. But, if such sifting helps ensure success for some, it does not solve the problems of others. Likewise, by and large, the successes, reported in this book, were made possible by the focusing of very significant funding. Because resource allocation is a zero-sum game, fewer dollars went to other neighborhoods, whose residents resented not sharing in the benefits of the targeted redevelopment (see, e.g., Chapters 5 and 8). Hence, working toward success can mean having to make tough equity decisions.

One must also ask: Success, but at what price? Marcuse notes how the per unit cost of rehabilitation can nearly equal that of new construction. Schill wonders if the improved quality of life for Lake Parc Place residents could be attained for others with less money through more flexible approaches such as the Gautreaux program. Ultimately, there is the question, raised by Vale, of whether we measure success by the rate at which its residents in assisted housing move on and up, or by the creation of safe, attractive, and stable environments where many of those who can afford to leave will choose to stay.

These are thorny questions. There are no easy answers. The chapters in this book reflect a broader lack of consensus in this regard. Nonetheless, the authors do identify outcomes that are clearly positive. Among them are the following:

- in New York, TIL produced 599 tenant-owned co-ops with a total of 12,922 apartments, with another 210 buildings with 4,128 units in the transitional phase of the program. The residents pay $250 for a unit, and its long-term affordability is assured. In comparison with tenants of city-owned buildings, the co-op residents express greater satisfaction with security, the provision of basic services, and management quality; in the co-ops there is less turnover, more informal socializing and assistance, and greater involvement in tenant activities, block associations, churches, and local government. Each TIL unit also saves the city $1,500 per year;
- surveys among the residents of Fidelis Way, Boston, in 1979, and its renamed renovation, Commonwealth, in 1993, show that the percentage

finding it a good place to raise kids increased from 19% to 78%, those feeling very safe in the project rose from 11% to 73%, and those without any complaints about maintenance jumped from 25% to 90%;

- after its renovation, Lake Parc Place in Chicago provides a much improved physical environment and has less crime; more than 80% of its residents say they are satisfied or very satisfied; very-low-income households, with a median income of $6,200, occupy 56% of the units for which there is a waiting list of 4,000;
- since its formation in 1981, the Cleveland Housing Network (CHN) has evolved into the nation's largest lease-purchase program; lease-purchasers pay $150-$250 for a housing unit, and 90% of them have incomes below 30% of the area median income;
- in Sandtown-Winchester, Baltimore, redevelopment to date has created more than 900 low-income housing units, the crime rate is down, health services have been provided to pregnant women and children, and new jobs have been created;
- in West Farms, the Bronx, 2,700 low-income apartments have been built or rehabilitated, local jobs have been created, crime is declining, a variety of social services are now available, access to health care has improved, new initiatives in education have gotten national recognition, an open space system is being developed, and the institutional fabric of the community is being rebuilt.

The authors provide greater detail of these and other positive outcomes. It is not possible to undertake a rigorous cost-benefit analysis of these cases, or to ascertain their underlying causality in a definitive way. However, it is still important to search for common patterns, associated with these successes. In line with other recent research (e.g., GAO, 1995), several themes emerge.

Ingredients of Success

A Comprehensive Approach

The successful redevelopment efforts documented in this book are predicated on the recognition that they must be based on a comprehensive approach. Although the scarcity of decent affordable housing was a problem in each case, these communities did not seek solutions through sectoral actions narrowly focused on housing alone. Instead, they realized that their problems were complex, consisting of multiple and interrelated issues. A comprehensive and focused approach has synergistic effects, which exceed the sum of partial efforts spread more widely. An incomplete list of redevelopment actions includes the following:

- housing rehabilitation
- new housing construction
- support for home ownership
- ownership training
- maintenance training
- collaborative decision making
- open space improvements
- project/neighborhood beautification
- individual and family counseling
- social services
- educational programs
- day care provision
- health care services
- recreational programs
- youth employment programs
- job training and creation
- crime reduction
- economic development
- teen parenting programs

Successfully addressing these issues requires operating simultaneously on several fronts. This demands a comprehensive approach that considers the effects that actions in one area will have elsewhere. For example, housing affordability is related to the availability of jobs with sufficient earning power, and the ability to fill those jobs is related to access to adequate schooling (cf. Gilderbloom & Wright, 1993; Shlay, 1995). Bratt, Simmons, and Vale, among others, describe the creation of jobs for neighborhood residents in new construction, rehabilitation of units, and the delivery of health care services as a source of generating local income. Such employment can occur within the context of a skills training program. Several chapters also report a variety of educational initiatives, such as programs for preschool activities, preparation for GED and college, as well as day care provision and youth employment schemes. Together, the chapters span a broad spectrum, from West Farms and

Sandtown-Winchester, with a full gamut of programs, to Lake Parc Place and CMHA with a more restricted range of interventions. However, in each case, the concerns go beyond merely providing housing. President Clinton's Urban Policy Report (HUD, 1995b) resonates this theme of a comprehensive approach to community revitalization, and the Community Empowerment Board, created in 1993, headed by the vice president and consisting of the heads of 15 federal agencies, has as its mission to support such approaches (see also HUD, 1996a).

A Community-Based Approach

Successful approaches typically focus on a specific community in a particular place, rather than target a certain category of households dispersed over a wide area. As noted, these approaches concern themselves with the whole of a community: its residents, housing stock, physical infrastructure, social service network, institutional organization, and so forth. The size of these communities varies considerably. Lake Parc Place is relatively small, comprising 282 units. Commonwealth is larger, 392 units, but still a single project. Although operating within a larger area, the Tenant Interim Lease Program, the Boston rehab programs, and the California Mutual Housing Association also tend to concentrate on specific sites. At the other end are West Farms (27 square blocks) and Sandtown-Winchester (72 square blocks), both of which, notwithstanding their scale, still provide a clearly delineated spatial focus for redevelopment. The "reinvention of HUD" includes an emphasis on place-based strategies that echo this community focus (Cisneros, 1995b, pp. 149-150).

It is important that the concern with whole communities does not produce an insular orientation. Krumholz states that Cleveland frequently undertakes projects on an ad hoc basis, without having an *overall* neighborhood revitalization strategy. Some projects are selected because of opportunity and price, rather than their "fit" into a larger plan. He urges that CHN activities be linked more strongly to a comprehensive city strategy for neighborhood improvement. This issue also relates to questions of equity in resource allocation, mentioned by Vale and Goetz, and is not restricted to Cleveland, but relevant to cities elsewhere as well.

Resident Participation

Community-based approaches emphasize meaningful participation by residents, empowering them with real responsibilities in all stages of the development process. For example, grassroots involvement was key to the success of TIL and the tenants' association played a central role in the Commonwealth redevelopment. The implementation of such participatory processes can be challenging. Krumholz notes how property owners and investors can take an active role on community development corporation (CDC) boards, skewing priorities away from affordable housing (cf. Varady, 1994; see also Bendick & Egan, 1993; Wiewel & Persky, 1994). He concludes that CDCs in Cleveland have not attained neighborhood empowerment, observing a gap between stated goals and real accomplishments. Richman and Heskin make the same point and propose CMHA as an alternative that combines local legitimacy with centralized services. They suggest an annual organizational audit and inclusion of a community representative as a board member as safeguards against isolation from the residents and the wider neighborhood. White and Saegert similarly caution against the risks of factionalization and the formation of cliques.

Participatory processes require skills in collaborative decision making, crisis mediation, and conflict resolution. TIL, CMHA, and others offer workshops to develop such skills among residents. Bratt also recommends that redevelopment activities be carried out through local nonprofits and resident associations to ensure congruence between needs and program goals. Based on the Boston experience, she concludes that the process of participation is as important as the product (see also Halpern, 1995; Medoff & Sklar, 1994; Tulloss, 1995).

Funding

The comprehensive nature of community redevelopment impinges on funding in major ways. To begin with, simultaneous coordinated action in several different realms is expensive. It takes a high level of financing to provide the foundation necessary for positive outcomes. This is readily illustrated by the cases in this book. The transformation of Fidelis Way into Commonwealth cost $31.5 million, translating into $80,000 per unit. Likewise, the make-over of Olander Homes into Lake Parc Place cost about $20 million, or $70,000 per unit. These amounts are not exceptional for projects of this type.[11] Obviously, the per unit price tags are not strictly comparable, because they include different things. Nevertheless, they do give an indication of how expensive it is to produce quality housing in major U.S. cities. Much higher levels of funding are required for more large-scale redevelopment efforts, as is clear from the chapters on West Farms and Sandtown-Winchester.

Adequate funding not only is needed for capital investment, but must also be available on a long-term basis for ongoing operations. Bratt, White and Saegert, Richman and Heskin, and Vale all stress the continuing need for funds to support maintenance and management tasks. This need is augmented by the aged housing stock occupied by many low-income households, necessitating more frequent and expensive upkeep and repairs. Chapters 4 and 9 both refer to the historical underfunding of property management in low-income communities in the United States. Long-term funding is also required to sustain the initial momentum of redevelopment. As noted below, turning around very difficult conditions on multiple fronts does not happen overnight. It takes staying power that cannot be fueled without money.

Funding is almost always obtained from multiple sources, each with its own specific application and accountability requirements. The chapter authors, Krumholz in particular, describe how these sources form complicated patchworks of loans and grants, often involving a variety of agencies at each of the three levels of government, as well as in the private sector, and in the nonprofit community. State and local government may also negotiate below-market interest rates with lenders and may leverage nonmonetary resources, such as land, infrastructure, or expertise (e.g., Cleveland and Baltimore make effective use of landbanking). There is no prototypical model after which communities can pattern themselves. Each situation is specific, with its own particular opportunities and constraints, producing uniquely configured financing arrangements. However, the cases in this book tend to share one feature: the significance of federal funding.[12]

Alternative sources of revenue have not been able to make up for the reduction in federal funding. This shortage is often compounded by the instability of local funding, as illustrated by the case of San Diego, where the recommended housing trust fund shrunk from $54 million to $13 million when the city council decided to divert the hotel room tax to help solve the city's fiscal woes, and Chicago, where an agreement with Presidential Towers to commit at least $14 million yielded merely $3.2 million when the project experienced financial setbacks (see Chapter 10). An advantage of using multiple funding sources is that it makes a project less dependent on any one sponsor, reducing its financial vulnerability.

Coalition Building

Success is not produced single-handedly. It takes the collaborative efforts of many groups. Initially, most partnerships work with a short-term agenda for a specific site. Subsequently, they may evolve into more enduring, institutionalized arrangements concerned with more long-term development goals (see, e.g., the chapters by Simmons, Goetz, Vale, and Bratt). Coalitions comprise a variety of constituencies in the community and representatives of various stakeholding offices at the city level and beyond. Participants from the private for-profit sector typically are lenders, developers, and builders. Others include foundations, charities, and unions. Frequently, there is an intermediary organization, such as the Enterprise Foundation (EF), the Local Initiatives Support Corporation (LISC), or the Neighborhood Reinvestment Corporation (NRC). These intermediaries often play a critical role. In Cleveland, for example, they helped with mobilizing equity and financial packaging, assisted with tax syndication, and legitimized CDCs for lenders. In one of her five recommendations, Bratt stresses the role of nonprofit intermediaries and their ability to insulate themselves from changes in the political landscape; other authors also bring out their significance in providing technical assistance.

Coalitions comprising multiple partners have greater organizational capacity and can better leverage available funds to raise additional support. The building of coalitions to gain political support is evident in the establishment of housing trust funds (Chapter 10). They also played a role in, for example, the selection for redevelopment of Commonwealth from among a larger number of potential sites (Chapter 5). A recent evaluation of the CDBG program in 61 cities indicates that coordination requires extra attention (Walker et al., 1995). Such coordination can be facilitated by umbrella organizations. In some ways, CMHA performs this function, providing technical assistance and fostering interrelationships among project-based MHAs, leaving other tasks to them. There is a somewhat similar division of labor between CHN and the CDCs in Cleveland. In New York, the Urban Homesteading Assistance Board set up a federation of co-ops that provides discounted insurance and legal services plans, a credit union, and bulk purchasing of fuel. TIL buildings have also organized themselves into neighborhood and citywide networks to reduce crime, exchange information, influence city policy, and offer peer support. White and Saegert suggest that a tenant-owned co-op service sector will minimize conflicts of self-interest among community organizations, government, and residents.

Management

Several of the studies in this book stress the importance of management. They report on various alternatives, including the privatization of public housing management (Chapter 6), the formation of tenant co-ops (Chapter 7), and resident control through a mutual housing association (Chapter 9). Regardless of their specific form, these diverse approaches all seek to ensure that management is accountable to the residents. Commonwealth tenants, for example, have little interest in taking on responsibility for day-to-day operations, but they can fire management with 30 days' notice. Other research suggests that the sense of control may be as much a function of the quality of relations with resident leaders as it is of opportunities for direct participation (Van Ryzin, 1994).

Aside from accountability, effective management also involves selectivity. Before tenants are admitted, they must first go through a careful screening process. Such screening can also occur through tenant associations or CDCs. Once accepted, tenants must abide by an agreed-on code of conduct. Effective management is firm in implementing the eviction of tenants who deal drugs, refuse to pay the rent, or otherwise violate their lease. Simmons also stresses the need for effective management to be professional in its day-to-day operations, attentive to the requirements of responsible property and asset management, and concerned with the social service needs of the residents. A broader view of management, incorporating these and related concerns, is echoed as well in other recent literature (e.g., Bratt, Keyes, Schwartz, & Vidal, 1994; Lane, 1995).

Physical Design

Successful redevelopment depends on much more than "bricks and mortar." However, the design of the housing and neighborhood environment does play an important role. Several chapters underscore this point. The Development Kit for the award-winning Commonwealth contained lengthy appendixes, including very detailed planning and design criteria. The plans called for grade changes in landscape to permit all private entrances to be entered from ground level; parking in small lots visible from the owner's apartment; play and picnic areas; and general avoidance of an institutional and repetitive appearance. Phipps Houses' Sunnyside in New York is still visited by architects and planners from around the world and owes its satisfaction among tenants in part to the quality of its design. West Farms residents attach great importance to the emerging open space system, which will not only create more play areas but also increase opportunities for informal socializing essential to the formation and maintenance of local support networks. Careful design also reduces opportunities for crime (cf. Cisneros, 1995a; Wekerle & Whitzman, 1995). Lake Parc Place derives its popularity in part from extra "amenities" such as closet doors, shower heads, miniblinds, ceiling fans, laundry rooms, and landscaped play areas. In contrast, the single most consistent complaint of TIL tenants is the low quality of the workmanship of government contractors. Evidence indicates that inadequate construction or rehabilitation can significantly raise the costs of property management (Bratt et al., 1994).

None of the chapters explicitly addresses issues of cultural diversity. However, current design codes and occupancy standards reflect mainstream values and may conflict with the cultural preferences of minorities. For example, the extended family lifestyle of many Mexican Americans is not well accommodated by conventional U.S. housing types and layout, nor are official regulations congruent with them (Pader, 1994).

In spite of the obvious significance of the physical environment, opinions differ over the priority it should receive. In Sandtown-Winchester, EF pushed hard for new construction, rehabilitation, and improvement of the physical infrastructure, whereas some residents wanted to give precedence to community building. This tension was also noticeable in West Farms and is not uncommon in similar situations elsewhere.

Professional Expertise

Each of the aforementioned points indicates the importance of specialized expertise. Sound planning, negotiating discounted mortgage loans, leveraging scarce resources, syndication, building productive partnerships, effective management, quality design—none of these can be accomplished by relying solely on the contributions of pro bono volunteers. The Commonwealth Tenants Association undertook a painstaking independent review of the renovation plans. It hired a licensed architect to construct scale models and with whom to consult. Similarly, White and Saegert describe how TIL tenants hired an outside expert to uncover shoddy work by the general contractor. They also note how co-op organizing has become more difficult as a result of the reduction in legal aid lawyers. Access to expert technical assistance is invaluable. In New York's TIL program, the city and the Urban Homesteading Assistance Board provide critical support in the form of technical and professional advice, services, and management training. In Cleveland, CHN is pivotal, for example, in the financing and managing of lease-purchase units and in training its members in prequalifying buyers and underwriting and credit practices. Cleveland has also effectively used the Community Reinvestment Act to reverse neighborhood disinvestment and channel loans worth hundreds of millions of dollars into city neighborhoods. Negotiating such agreements demands close familiarity with the requirements of the law along with political savvy. In West Farms, Phipps Community Development Corporation was instrumental in establishing and coordinating an array of programs and activities that helped rebuild the local community, whereas CMHA's mission revolves around extending technical assistance in support of resident control. Simmons's experience as CEO of Phipps Houses, the nation's oldest and one of its largest nonprofit housing providers, leads her to insist that the professionalization and continuity of urban redevelopment demand a commitment to well-trained and competitively paid staff.

Time

For the communities that reversed their misfortunes, success did not come overnight. The unfolding of resident participation takes time, as does the securing of funding, the developing of a social service network, and myriad other tasks. In some cases, it took 15 years or more before a turnaround became apparent. CHN was founded in 1981, as the neighborhood advocacy groups of the 1970s evolved to become more development oriented, producing a small but committed and competent cadre now making up an effective network of CDC directors and staff, city hall officials, politicians, and lending officers. The TIL program was begun in 1978, but it took until the late 1980s before the city seriously committed to it. Phipps Houses operated for 20 years in West Farms before its efforts more fully began to come to fruition.

Setbacks do occur. Again, West Farms offers case material, such as contractor fraud in the construction of Lambert Houses, the demise of both TIP and the Bronx Park South Management Corporation, and opposition from corrupt and hostile local politicians. These episodes illustrate the obvious: Progress does not follow a linear path. However, it is possible to learn lessons along the way. An instructive analysis comes from Bratt, who shows how the success of Boston's current Demo Dispo Program builds on the experience with previous rehab programs going back to 1968. Based on this analysis, one of her recommendations is that the redevelopment schedule should be sufficiently generous to allow time for the residents and community leaders to understand fully and embrace the program and, if necessary, to modify it to better meet local needs.

Identifying characteristics of success is one thing. However, learning from them is quite another. What lessons do the cases in this book hold? How can they inform current and prospective policies? These questions concern the transferability of successful experiences. They are the subject of the next section.

Transferability of Success

To the extent and in the ways that the U.S. challenges of affordable housing and urban redevelopment are context specific, study of the experience of selected U.S. cities may provide an apposite and readily accessible basis for drawing lessons. City agencies typically comprise functionally specialized divisions, which logically search for insights from their counterparts in different jurisdictions facing similar problems. The National Low Income Housing Information Service, located in Washington, D.C., aims to facilitate such interlocal learning. HUD's Clearing Houses have a broader but similarly supportive mission. Furthermore, each year since 1989, the Agency Awards of Excellence Program in Housing and Community Development of the National Association of Housing and Redevelopment Officials has selected exemplary local programs and projects that may serve as models for communities elsewhere.[13] The Ford Foundation's annual Innovations in State and Local Government Awards Program fulfills a comparable role. Since its inception in 1985, 80% of the award-winning programs have been replicated or significantly expanded (Ford Foundation, 1994). The goal of identifying successful models also underlies the worldwide contest for "best practices," organized in 1996 by the U.N. Centre for Human Settlements as part of the much heralded Habitat II conference (see also HUD, 1996b).

Why do cities look for lessons from other cities? Dissatisfaction with the current situation is the primary motivating factor influencing the readiness to look for answers elsewhere. Whether dissatisfaction becomes a significant factor depends, in large part, on the degree of consensus among decision makers and experts that there is inescapable evidence of policy failure, electoral pressures, and the prevalence of political values favoring change (Rose, 1993).

If it can be determined that a search for lessons is ideologically desirable and technically practicable, where does one look? Resources are clearly a consideration. Cities in fiscal crisis are unlikely to try an approach that is working elsewhere but requires a strong revenue base. Geographical propinquity also plays a role. Ceteris paribus, learning from nearby communities is cheaper and more convenient. Furthermore, ideological similarity carries weight.[14]

Readiness to learn is one thing. Translating this into practice is quite another. How can the formulation of policies in one place be informed by the experience of another place? A recent comparative assessment of neighborhood regeneration programs in nine countries ends with a pessimistic note, suggesting that we lack the knowledge that would permit us to specify cause-and-effect models (Alterman, 1995). However, Rose (1993) distinguishes alternative ways of drawing lessons, including copying (enacting a more or less intact program already in existence elsewhere), adaptation (adjusting a program for contextual differences), synthesis (combining elements from different sources into a new program), and inspiration (using programs elsewhere as a stimulus for innovation). In a comparison of approaches to housing and neighborhood renewal in Britain and Sweden, Elander (1995) adopts network theory as a conceptual framework for the analysis of implementation processes. Some of his findings parallel those from the case studies in this book (e.g., concerning the prominent role of tenant organization, community participation, intergovernmental cooperation, and public-private partnerships). Andersen (1995) looks for lessons Europe may learn from cycles of decay and renewal in the United States, but does not go beyond observing a general pattern of differences in government intervention.

Whatever the approach taken, lesson drawing is different from the diffusion of successful prototypes, which assumes replication through a simple technocratic process and discounts differences in the political and economic context. Instead, effective lesson drawing requires prospective evaluation. Prospective evaluation combines empirical evidence about how a program operates in one place with hypotheses about the likely effects of a similar program in another place. Its purpose is to reduce failure through anticipatory feedback. It starts with a comparison between the actual conditions associated with a program already in effect and the presence or absence of these conditions elsewhere. The critical step is applying knowledge about what works somewhere else, and why, to one's own circumstances (Rose, 1993).

It follows, then, that no universal magic formula can emerge from the experience of any one place. Lesson drawing requires specification. It will not produce a single meta-model. Prospective evaluation must occur with reference to the distinct characteristics of a particular city. Criteria exist that can guide such assessment. Rose (1993) suggests three factors that identify qualifying conditions for lesson drawing and influence the fungibility of programs.

First, the fewer its unique elements, the more fungible a program is. In their specificity, the approaches described in this book are obviously unique. However, at another level, they all seek to provide affordable housing and a suitable living environment in declining city neighborhoods with certain common economic, physical, and social characteristics. In that sense, the programmatic concerns are not one of a kind but occurring in a limited number of places. Hence, these cases are not unique, but context dependent, and insofar as different cities share relevant contexts, in principle, their programs can transfer.

Second, the more substitutable its institutions of program delivery, the more fungible a program is. Adopting a program from elsewhere requires an institutional capacity to do so. However, institutions need not be identical. For example, effective management practices emerge as an important theme from the case studies. But, as noted above, effective management can be provided by different organizational forms. Likewise, technical assistance and professional expertise in redevelopment are critical. Again, the case studies show the variety of mechanisms through which they can be provided. Similarly, whereas those doing the fund-raising, and the specific sources tapped, vary from place to place, the principles of coalition formation and of leveraging multiple sponsors apply more broadly.

Third, the greater the equivalence of resources, the more fungible a program is. Implementing programs requires resources. Among them are a legislative mandate, qualified personnel, and money to pay for salaries, subsidies, land, materials, and so on. U.S. cities all operate under the same federal laws. Although state and local laws may differ, the problems of adjusting existing legal frameworks or creating new ones are political, rather than technical. This is clear from, for example, housing trust funds, whose principles are well understood, but whose adoption is very much a function of political desirability (see Chapter 10). Nor is there evidence indicating that U.S. cities differ significantly in the expertise required to carry out successful redevelopment.

Resources do vary greatly. Some cities have less built-up land than others. Some have more money than others. However, in the latter case, the issue may be less a matter of adequate revenues, and more a matter of assigning a low political priority to redevelopment when making inevitable trade-offs, or a reluctance to raise taxes or charge impact fees. Insofar as money is a real issue, it may be better to redirect lesson drawing, rather than to abandon it altogether.

These criteria cannot specify a general panacea for the woes of our inner cities. However, they can help guide searches for locally specific solutions. These searches will be more productive if they are based on the recognition that to replicate an approach or to transfer a program successfully, the circumstances must be functionally equivalent, but that it is not essential that they be identical. Furthermore, fungibility is often not an either/or question but a matter of degree.

Conclusion

Nearly all case studies in this book document and analyze the redevelopment of declining urban neighborhoods in large cities. From these studies emerge the broad contours and features of a successful approach. This approach is generally characterized by being comprehensive and focused. It is based on substantial and sustained funding from multiple sources and relies on strong coalitions consisting of partners from the public, private, nonprofit, and community sectors. It includes meaningful resident participation, but also requires technical assistance and professional expertise. It uses careful physical design and effective management to assure adequate housing. And, finally, it takes time.

Do all of the cases neatly fit this general model? No, they do not, and it is perhaps instructive that the model applies less to those cases whose success is more qualified.

Housing trust funds are perhaps least conforming. They are an attempt by state, county, and local governments to create a dedicated source of revenue for affordable housing in response to federal cutbacks. The revenues they generate provide welcome support, but they are not inherently stable (e.g., San Diego, Chicago). They are also limited in scale (e.g., Cleveland) and not sufficient to make up for the federal shortfall. Their allocation is not usually concentrated and is directed predominantly to housing rather than more broadly to redevelopment. In this sense, trust funds fulfill an important but limited function.

The general model is not a perfect fit for CMHA either. CMHA, too, is closely oriented to housing. Although it deals with specific sites, it operates statewide. The TIL program and CHN have a similar umbrella function with multiple foci, but offer a somewhat broader array of services. CMHA has only a short history. The jury is necessarily still out on its eventual success. However, its experience to date suggests the effectiveness of an approach that contains some but not all of the model's characteristics.

MINCS's Lake Parc Place is smallest in scale, and the scope of its redevelopment was also more immediately restricted to housing per se. Although it is deemed a success, its cost-effectiveness has been questioned. Its mixed-income population is clearly not a necessary condition for success, considering, for example, Commonwealth, which appears at least equally successful but has an exclusively low-income population. MINCS was adopted only by Chicago, and even there Lake Parc Place is only a partial implementation. The recent federal takeover of the city's housing authority casts further doubt on the program's continuation. However, just as the absence of the dog's barking was important information to Sherlock Holmes in his efforts to find the murderer, so also can the absence of compelling evidence be useful in identifying the limitations of any given approach.

The case studies are a font of information. Some things can be learned from what works well and may be replicable; other things can be learned from what did not work and should be avoided. Some of these lessons can be found in the generalities the cases have in common. Other lessons can be found in their rich detail. However, readers looking for an open sesame will look in vain. There is no single meta-model. Hence, the lessons we draw will necessarily differ according to our individual circumstances.

NOTES

1. See Link, Phelan, and Bresnahan (1994) for a discussion of the flawed methods that have produced downwardly biased estimates.

2. *Extremely low* incomes are below 25% of the area median income (AMI); *very low* incomes are between 25% and 50% of AMI, and *low* incomes are between 50% and 80% of AMI.

3. Recently, the American Bar Association passed a resolution that calls on state and federal lawmakers to protect victims of domestic abuse from insurance discrimination. HR 1201, introduced in Congress by Ron Wyden on March 10, 1995, aims to counter this practice.

4. The Fair Housing Amendments Act of 1988 made discrimination on the basis of family status illegal. However, several local studies indicate the continued existence of discriminatory rental practices that exclude families with children (e.g., Galster, in press).

5. Much of the literature assumes similar effects on cities in lower tiers of the world's urban hierarchy. However, empirical documentation in support of this assumption remains scarce.

6. Others have rejected the term *underclass* because its conceptual weaknesses limit its explanatory power and because its behavioral connotations obscure its underlying structural causes (see, e.g., Lee, 1994).

7. Nord's (1990) analysis shows how decentralization produced inequality in the distribution of program benefits in Sweden.

8. By the same token, to the extent and in the ways that cities differ from each other in relevant circumstances, it becomes necessary to specify local situations when examining the effects of global restructuring and national policy shifts (Burgers, in press).

9. The panel comprised members of the editorial boards of leading academic journals concerned with urban affairs and economic development and members of the Executive Boards of two prominent economic development organizations. The index of urban distress was created from indicators identical or very similar to those listed in note 10.

10. Included were data on changes in the unemployment rate, median household income, proportion of the population below the poverty line, per capita income, and labor force participation.

11. The per unit cost was considerably higher in two recent mixed-income projects, combining residential and various commercial functions. Harbor Point, an award-winning (mixed-income) renovation of the infamous Columbia Point in Boston, completed in 1991, cost $250 million, or $195,000 per unit, whereas Tent City, another successful project in that city, cost $38 million, or $140,000 per unit.

12. In line with this, a recent review of comprehensive redevelopment efforts in Boston, Detroit, Washington, D.C., and Pasadena, California, found that public funding made up between 30% and 60% of the budget, much of it coming from CDBG or CSBG (GAO, 1995). See Wallace (1995) for a good discussion of (mostly) federal government support for low-income housing within a broader framework of financing the affordability gap.

13. The list of award-winning projects in 1995 includes an index on nearly 70 topics. For more information, contact NAHRO, 1520 18th Street, N.W., Washington, DC 20036-1811. Ph.: (202) 429-2960. NLIHIS can be contacted at 1012 14th Street, N.W., Washington, DC 20005. Ph.: (202) 662-1530.

14. For example, a review of six labor and welfare laws found that 40 of the 48 continental states consistently favor programs on either the left or the right of the political spectrum (Robertson, 1991, cited in Rose, 1993, pp. 102-103).

REFERENCES

Abramson, A. J., Tobin, M. S., & Vander Goot, M. R. (1995). The changing geography of metropolitan opportunity: The segregation of the poor in U.S. metropolitan areas, 1970 to 1990. *Housing Policy Debate, 6,* 45-72.

Alterman, R. (1995). A comparative view of neighborhood regeneration programs in nine countries. *Urban Affairs Review, 30,* 749-765.

Andersen, H. S. (1995). Explanations of decay and renewal in the housing market: What can Europe learn from American research? *Netherlands Journal of Housing and the Built Environment, 10*(1), 65-85.

Bendick, M., Jr., & Egan, M. L. (1993, August). Linking business development and community development in inner cities. *Journal of Planning Literature, 8*(1), 3-19

Bratt, R. G., Keyes, L. C., Schwartz, A., & Vidal, A. C. (1994). *Confronting the management challenge: Affordable housing in the nonprofit sector.* New York: Community Development Research Center, Graduate School of Management and Urban Policy, New School for Social Research.

Burchell, R. W., & Listokin, D. (1995). Influences on United States housing policy. *Housing Policy Debate, 3,* 559-617.

Burgers, J. (in press). No polarization in Dutch cities? Inequality in a corporatist society. *Urban Studies.*

Burgers, J., & Kalb, D. (1995). Economic restructuring and local consequences. *Netherlands Journal of Housing and the Built Environment, 10,* 127-139.

Burt, M. R. (1995, July). Critical factors in counting the homeless: An invited commentary. *American Journal of Orthopsychiatry, 65,* 334-339.

Cerny, P. G. (1994). The dynamics of financial globalization: Technology, market structure, and policy response. *Policy Sciences, 27,* 319-342.

Chicago Title and Trust Company. (1996). *Who's buying homes in America?* Chicago: Chicago Title and Trust Family of Title Insurers' 20th Annual Survey of Recent Home Buyers.

Cisneros, Henry G. (1993, December 5). The lonely death on my doorstep. *Washington Post,* sec. C, p. 1.

Cisneros, Henry G. (1995a, February). *Defensible space: Deterring crime and building community.* Washington, DC: U.S. Department of Housing and Urban Development.

Cisneros, Henry G. (1995b, September). Legacy for a reinvented HUD: Charting a new course in changing and demanding times. *Cityscape: A Journal of Policy Development and Research, 1,* 145-152.

Clarke, S. E. (1989). The political implications of fiscal policy changes. In S. E. Clarke (Ed.), *Urban innovation and autonomy* (pp. 236-251). Newbury Park, CA: Sage.

Clarke, S. E. (1993). The new localism: Local politics in a global era. In E. G. Goetz & S. E. Clarke (Eds.), *The new localism: Comparative urban politics in a global era* (pp. 1-21). Newbury Park, CA: Sage.

Davis, J. E. (Ed.). (1994). *The affordable city; Toward a third sector housing policy.* Philadelphia: Temple University Press.

Dreier, P., & Atlas, J. (1996). *U.S. housing policy at the crossroads: A progressive agenda to rebuild the housing constituency.* Los Angeles: Occidental College, International & Public Affairs Center.

Elander, I. (1995). Policy networks and housing regeneration in England and Sweden. *Urban Studies, 32,* 913-934.

Fainstein, S. S. (1994). *The city builders. Property, politics and planning in London and New York.* Cambridge: Basil Blackwell.

Ford Foundation. (1994). *Ford Foundation report.* New York: Author.

Frieden, B. J., & Sagalyn, L. B. (1989). *Downtown, Inc.: How America rebuilds cities.* Cambridge: MIT Press.

Galster, G. (in press). Discrimination. In W. van Vliet– – (Ed.), *Encyclopedia of housing.* New York: Garland.

Gilderbloom, J., & Wright, M. (1993). Empowerment strategies for a low income African American neighborhood. *Harvard Journal of African American Public Policy, 2,* 77-95.

Goetz, E. G. (1993). *Shelter burden: Local politics and progressive housing policy.* Philadelphia: Temple University Press.

Goetz, E. G., & Clarke, S. E. (Eds.). (1993). *The new localism: Comparative urban politics in a global era.* Newbury Park/London/New Delhi: Sage.

Halpern, R. (1995). *Rebuilding the inner city: A history of neighborhood initiatives to address poverty in the United States.* New York: Columbia University Press.

Harloe, M., Marcuse, P., & Smith, N. (1992). Housing for people, housing for profits. In S. S. Fainstein, I. Gordon, & M. Harloe (Eds.), *Divided cities* (pp. 175-202). Oxford, UK and Cambridge, MA: Basil Blackwell.

Heisler, B. S. (1994). Housing policy and the underclass: The United Kingdom, Germany, and the Netherlands. *Journal of Urban Affairs, 16,* 203-220.

Henwood, D. (1996, January 22). America: Still world capital of inequality. *Left Business Observer, 71,* 4-5.

Hopper, K. (1995, July). Definitional quandaries and other hazards in counting the homeless: An invited commentary. *American Journal of Orthopsychiatry, 65,* 340-346.

Ihlanfeldt, K. (1994, August). The spatial mismatch between jobs and residential locations within urban areas. *Cityscape: A Journal of Policy Development and Research, 1,* 219-244.

Jargowsky, P. A. (1994). Ghetto poverty among Blacks in the 1990s. *Journal of Policy Analysis and Management, 13,* 288-310.

Joint Center for Housing Studies. (1994). *The state of the nation's housing: 1994.* Cambridge, MA: Joint Center for Housing Studies of Harvard University.

Kasarda, J. (1993). Inner-city concentrated poverty and neighborhood distress: 1970 to 1990. *Housing Policy Debate, 4,* 253-302.

Kozol, J. (1988). *Rachel and her children.* New York: Fawcett Columbine.

Ladd, E. C. (Ed.). (1995). *America at the polls.* Roper Center, University of Connecticut, Storrs, CN.

Ladd, H. F. (1982, May). Equal credit opportunity: Women and mortgage credit [New York and California]. *American Economic Review, 72,* 166-170.

Lane, V. (1995). Best management practices in U.S. public housing. *Housing Policy Debate, 6,* 867-904.

Lee, P. (1994). Housing and spatial deprivation: Relocating the underclass and the urban poor. *Urban Studies, 31,* 1191-1209.

Link, B. G., Phelan, M., & Bresnahan, M. (1994). Lifetime and five-year prevalence of homelessness in the United States. *American Journal of Public Health, 84,* 1907-1912.

Logan, J. R., & Swanstrom, T. (Eds.). (1990). *Beyond the city limits.* Philadelphia: Temple University Press.

Marans, R. W., & Colten, M. E. (1985). United States rental housing practices affecting families with children: Hard times for youth. In W. van Vliet– –, E. Huttman, & S. F. Fava (Eds.), *Housing needs and policy approaches: Trends in thirteen countries* (pp. 41-58). Durham, NC: Duke University Press.

Marcuse, P. (1990). Homelessness and housing policy. In C. Caton (Ed.), *Homeless in America.* New York/Oxford: Oxford University Press.

Massey, D. S., & Eggers, M. L. (1993). The spatial concentration of affluence and poverty during the 1970s. *Urban Affairs Quarterly, 29,* 299-315.

Massey, D. S., Gross, A. B., & Shibuya, K. (1994, June). Migration, segregation, and the geographic concentration of poverty. *American Sociological Review, 59,* 425-445.

Medoff, P., & Sklar, H. (1994). *Streets of hope: The fall and rise of an urban neighborhood.* Boston: South End.

Myers, S. L., Jr., & Chan, T. (1995, September). Racial discrimination in housing markets: Accounting for credit risk. *Social Science Quarterly, 76,* 543-561.

Nord, L. (1990). National housing policy and local politics in Sweden. In W. van Vliet– – & J. van Weesep (Eds.), *Government and housing: Developments in seven countries* (pp. 67-79). Newbury Park/London/New Delhi: Sage.

O'Connell, V. (1994, February). How to fight lenders' age and sex bias. *Money, 23,* 42-44.

Pader, E.-J. (1994). Spatial relations and housing policy: Regulations that discriminate against Mexican-origin households. *Journal of Planning Education and Research, 13,* 119-135.

Peterson, P. (1981). *City limits.* Chicago: University of Chicago Press.

Ritzdorf, M. (1986). Adults only: Children and American city planning. *Children's Environments Quarterly, 3*(4), 26-33.

Robertson, D. B. (1991). Political conflict and lesson-drawing. *Journal of Public Policy, 2*(1), 55-78.

Rohe, W. M., & Stegman, M. A. (1990). *Public housing homeownership demonstration assessment.* Washington, DC: U.S. Department of Housing and Urban Development.

Rose, R. (1993). *Lesson-drawing in public policy.* Chatham, NJ: Chatham House.

Sassen, S. (1991). *The global city: New York, London, Tokyo.* Princeton, NJ: Princeton University Press.

Sassen, S. (1994). *Cities in a world economy.* Thousand Oaks, CA: Pine Forge.

Savitch, H. V. (1995). The emergence of delocalized cities. *Urban Affairs Review, 31,* 137-142.

Shlay, A. B. (1995). Housing in the broader context in the United States. *Housing Policy Debate, 6,* 695-720.

Stoesz, D., & Karger, H. J. (1993). Deconstructing welfare: The Reagan legacy and the welfare state. *Social Work, 38,* 619-628.

Stone, M. E. (1993). *Shelter poverty.* Philadelphia: Temple University Press.

Taylor, J., & Jureidini, R. (1994, June). The implicit male norm in Australian housing finance. *Journal of Economic Issues, 28,* 543.

Tulloss, J. K. (1995, March). Citizen participation in Boston's development policy: The political economy of participation. *Urban Affairs Review, 30,* 514-537.

U.S. Bureau of the Census. (1995). *Statistical abstract of the United States 1994* (114th ed.). Washington, DC: Author.

U.S. Department of Housing and Urban Development. (1995a). *American housing survey for the United States in 1993* (Current Housing Reports H150/93). Washington, DC: Author.

U.S. Department of Housing and Urban Development. (1995b, July). *Empowerment: A new covenant with America's communities* (President Clinton's National Urban Policy Report). Washington, DC: Government Printing Office.

U.S. Department of Housing and Urban Development. (1995c). *HUD reinvention: From blueprint to action* (plus Summary). Washington, DC: Author.

U.S. Department of Housing and Urban Development. (1996a). *Beyond shelter: Building communities of opportunity.* Washington, DC: Author.

U.S. Department of Housing and Urban Development. (1996b). *Communities at work: Addressing the urban challenge.* National Excellence Awards for the City Summit. Washington, DC: Author.

U.S. General Accounting Office. (1995, February). *Community development: Comprehensive approaches address multiple needs but are challenging to implement* (GAO/RCED/HEHS-95-69). Washington, DC: Author.

Van Ryzin, G. G. (1994). Residents' sense of control and ownership in a mutual housing association. *Journal of Urban Affairs, 16,* 241-253.

van Vliet--, W. (1990). The privatization and decentralization of housing. In W. van Vliet-- & J. van Weesep (Eds.), *Government and housing: Developments in seven countries* (pp. 9-24). Newbury Park, London, New Delhi: Sage.

Varady, D. P. (1994). Middle-income housing programmes in American cities. *Urban Studies, 31,* 1345-1366.

Vidal, A. C. (1995). Reintegrating disadvantaged communities into the fabric of urban life: The role of community development. *Housing Policy Debate, 6,* 169-230.

Walker, C., et al. (1995, Winter/Spring). An assessment of the nation's first block grants. In *Policy and Research Report.* Washington, DC: Urban Institute.

Wallace, J. E. (1995). Financing affordable housing in the United States. *Housing Policy Debate, 6,* 785-814.

Wekerle, G. R., & Whitzman, C. (1995). *Safe cities: Guidelines for planning, design, and management.* New York: Van Nostrand Reinhold.

Wexler, R. (1995). Urban violence and community revitalization: The case of Sandtown-Winchester. *Planning Forum, 1,* 7-18.

Wiewel, W., & Persky, J. (1994). Urban productivity and the neighborhoods: The case for a federal neighborhood strategy. *Environment and Planning C: Government and Policy, 12,* 473-483.

Wilson, P. A. (1995). Embracing locality in local economic development. *Urban Studies, 32,* 645-658.

Wilson, W. J. (1987). *The truly disadvantaged: The inner city, the underclass and public policy.* Chicago: University of Chicago Press.

Wolff, E. N. (1995). *Top heavy: A study of the increasing inequality of wealth in America.* New York: Twentieth Century Fund.

Wolman, H. L., Ford, C. C., III, & Hill, E. (1994). Evaluating the success of urban success stories. *Urban Studies, 31,* 835-850.

Wright, J. D., & Devine, J. A. (1995, July). Housing dynamics of the homeless: Implications for a count. *American Journal of Orthopsychiatry, 65,* 320-329.

Wrightson, M. (1986). Interlocal cooperation and urban problems: Lessons for the new federalism. *Urban Affairs Quarterly, 22,* 261-275.

Yinger, J. (1995). *Closed doors, opportunities lost: The continuing costs of housing discrimination.* New York: Russell Sage.

Index

About the Contributors

Rachel G. Bratt is Associate Professor and Chair of the Department of Urban and Environmental Policy at Tufts University, specializing in housing and community development. She is a member of the Multifamily Advisory Committee of the Massachusetts Housing Finance Agency. She received a Ph.D. in Urban Studies and Planning from the Massachusetts Institute of Technology and is the author of *Rebuilding a Low-Income Housing Policy* and coeditor of *Critical Perspectives on Housing*.

Mary E. Brooks has worked as a low-income housing advocate for over 20 years. The majority of her work has involved applied research on housing, land use and zoning, community development, and civil rights issues for nonprofit organizations and government agencies. Currently, she directs the Housing Trust Fund Project for the Center for Community Change, a nonprofit organization providing technical assistance and other services to nonprofit organizations in low-income communities nationwide. A quarterly newsletter with information on the development and implementation of housing trust funds, other reports, and technical assistance are available from the center.

Susan S. Fainstein is Professor of Urban Planning and Policy Development at Rutgers University. She is coeditor of *Divided Cities: New York and London in the Contemporary World* and author of *The City Builders: Property, Planning and Politics in London and New York.* She has written extensively on issues of redevelopment, comparative public policy, and urban social movements.

Edward G. Goetz is Associate Professor in the Housing Program at the University of Minnesota. His research interests include local housing policy and politics, community-based housing development, local economic development policy, and homelessness. His book *Shelter Burden: Local Politics and Progressive HousingPolicy* was published in 1993. He was coeditor, with Susan Clarke, of *The New Localism: Local Politics in a Global Era* (1993) and has published articles on local housing policy in *Urban Affairs Quarterly, Journal of the American Planning Association, International Journal of Urban and Regional Research,* and *Journal of Urban Affairs.* He is currently working on the neighborhood-based politics of subsidized housing.

Allan David Heskin was on the staff of the National Housing Law Project and is currently Professor in the Urban Planning Program at the University of California, Los Angeles. A major focus of his research is resident-controlled affordable housing, particularly limited equity cooperatives. He wrote *Tenants and the American Dream* and, most recently, *The Struggle for Community,* which has won four national awards. His book *The Hidden History of Housing Cooperatives* is forthcoming. He is also an active practitioner in community development in southern California while serving as President of the California Mutual Housing Association.

Norman Krumholz is Professor in the College of Urban Affairs, Cleveland State University. He previously served as an urban planner in Ithaca, New York, and Pittsburgh, Pennsylvania, and as Planning Director of Cleveland from 1969 to 1979. He has published in many professional journals, including *Journal of the American Planning Association, Journal of Planning Education and Research,* and *Journal of Urban Affairs,* and he has written chapters for nine books. His most recent books are *Making Equity Planning Work: Leadership in the Public Sector* (with John Forester) and *Reinventing Cities: Equity Planners Tell Their Stories.* He received the APA award for Distinguished Leadership in 1990 and was awarded the Prize of Rome in 1987 by the American Academy in Rome.

Peter Marcuse, a lawyer and urban planner, has been involved in urban policy decisions for many years. He was a member of Waterbury, Connecticut's, City Planning Commission, later President of the Los Angeles Planning Commission, and has been a member of Community Board 9 in Manhattan and cochair of its housing committee. He was in the private practice of law in Waterbury for over 20 years. Since 1975, he has been Professor of Urban Planning at Columbia University in New York City. He has long-standing interests in comparative housing and planning policies. His numerous publications include articles on housing policy, red-lining, racial segregation, urban divisions, New York City's planning history, and the history of housing.

Mary K. Nenno was a Visiting Fellow at the Urban Institute, Washington, D.C. (1992-1996) and is the author of *Ending the Stalemate: Moving Housing and Urban Development Into the Mainstream of America's Future* (1996). She has been involved in the housing and urban development field for over 40 years, beginning as a staff member of the Buffalo Municipal Housing Authority and then serving in various positions on the staff of the National Association of Housing and Redevelopment Officials in Washington, D.C., from 1960 to 1991. She is a member of the Fannie Mae Housing Policy Research Advisory Board and has served as a consultant to the John F. Kennedy School of Government of Harvard University and the Enterprise Foundation. Her special expertise concerns housing policy at federal, state, and local levels; low-income housing for the elderly; metropolitan housing; and linking housing, welfare, and human services. She has directed numerous research projects funded by the federal government and foundations and is the author or editor of 17 books or monographs and over 50 articles.

Neal Richman teaches at the University of California, Los Angeles, Department of Urban Planning in the School of Public Policy and Social Research on such topics as real estate development, planning theory, and professional practice. He received his doctoral degree for comparative housing research from the Department of Development and Planning at the University of Aalborg in Denmark, where he continues to teach a module on planning ethics each year. He has 15 years of experience in affordable housing development, contributing to the rehabilitation and construction of more than 1,000 dwelling units in southern California. His clients now include nonprofit organizations in Western Europe and South Africa, and he remains a consultant with the California Mutual Housing Association.

Susan Saegert is Professor of Environmental Psychology and Director of the Center for Human Environments at the City University of New York Graduate School and University Center. She directs the work of the Housing Environments Research Group, which focuses on the relationship of housing to human development and the social fabric of communities.

Michael H. Schill is Professor of Law and Urban Planning at New York University. He directs the Center for Real Estate and Urban Policy at NYU. He has written a book and several articles on various aspects of housing policy, finance, and discrimination. His current research projects include analyses of the determinants of housing abandonment in New York City, the effect of public housing on neighborhood poverty rates, and a study of enforcement of laws prohibiting discrimination in the housing market.

Lynda Simmons is President Emerita of the Phipps Houses (PH) Group of New York City, America's oldest (1905) public service real estate firm, which she built from 60 staff and 500 apartments owned in 1969 to 500 staff and 3,700 units owned at her retirement in 1993. During her tenure, PH created nine communities in new and rehabilitated buildings, ranging from 39 apartments for homeless mothers and children in the South Bronx to 894 apartments in midtown Manhattan. In 1993, PH owned nearly a half-billion dollars' worth of real estate, and managed another near-half billion worth for other not-for-profit owners. In 1974, she developed PH's human services/community development capacity through the Phipps Community Development Corporation. She still serves on the PH and PCDC Boards, as well as that of the Environmental Simulation Center of the New School for Social Research.

Holly Sklar is the author, with Peter Medoff, of *Streets of Hope: The Fall and Rise of an Urban Neighborhood* (1994), which explores the Dudley Street Neighborhood Initiative, an innovative and successful resident-driven community revitalization project located in Boston. Her latest book is *Chaos or Community? Seeking Solutions, Not Scapegoats for Bad Economics* (1995). Among her earlier works is *Poverty in the American Dream* (1983), coauthored with Barbara Ehrenreich and Karin Stallard. She has contributed to numerous anthologies, newspapers, and magazines such as the *Philadelphia Inquirer, Kansas City Star, Miami Herald, USA Today, Z Magazine,* and *The Nation.*

Lawrence J. Vale is Associate Professor of Urban Studies and Planning at Massachusetts Institute of Technology, where he has taught since 1988. He is the author of two books on the design and policy underpinnings of government-subsidized environments. His *Architecture, Power, and National Identity* (1992) received the 1994 Spiro Kostof Book Award for Architecture and Urbanism from the Society of Architectural Historians. His current research, aided by a Guggenheim Fellowship, assesses recent efforts to revitalize public housing developments, to be published in a book about the relationship between public housing design and policy. In 1991-1992, he served as a consultant to the National Commission on

Severely Distressed Public Housing and, in 1995, was guest editor for a theme issue of *Journal of Architectural and Planning Research* titled "Public Housing Transformations."

Willem van Vliet– – became immersed in housing problems by birth, below sea level in an aporphyrogenic bunker in the postwar shortage-ridden Netherlands. He has an abiding interest in heather morning glory and rock gardening and is affiliated with CU. He is in possession of an uncertified but authentic lunatic streak, evinced, inter alia, by his assenting to the role of editor of the forthcoming *Encyclopedia of Housing.*

Avis C. Vidal is Director of the Community Development Research Center and Associate Professor of Urban Policy at the New School for Social Research. She has studied community-based development organizations extensively. Her writings on this topic include *Confronting the Management Challenge: Affordable Housing in the Nonprofit Sector and Rebuilding Communities: A National Study of Urban Community Development Corporations.* She recently placed the work of CDCs in a broader economic development and public policy context in "Reintegrating Disadvantaged Communities Into the Fabric of Urban Life: The Role of Community Development" in *Housing Policy Debate.* Prior to joining the New School, she spent 6 years on the faculty of the Kennedy School of Government at Harvard University and 2 years on the Legislative and Urban Policy Staff at HUD, earning the Merit Award for Outstanding Service.

Andrew White is the editor of *City Limits* magazine, an urban affairs monthly in New York City, and executive director of City Limits Community Information Service, a nonprofit organization dedicated to the dissemination of information about the revitalization of New York's low- and moderate-income communities. He is a graduate of the Columbia University Graduate School of Journalism. His work has appeared in the *Village Voice, Metropolis,* the *San Francisco Chronicle,* and elsewhere.